Deep Learning in Diabetes Mellitus Detection and Diagnosis

Deep Learning in Diabetes Mellitus Detection and Diagnosis focuses on deep learning-based approaches in the field of diabetes mellitus detection and diagnosis, including preprocessing techniques that are an essential part of this subject. This is the first book of its kind to cover deep learning-based approaches in the specific field of diabetes mellitus. This book includes a detailed introductory overview as well as chapters on current applications, preprocessing of data using deep learning, deep learning techniques, complexity, challenges, and future directions. It will be of great interest to researchers and professionals working on diabetes mellitus as well as general medical applications of machine learning.

Features:

- Highlights how the use of deep neural networks-based applications can address new questions and protocols, as well as improve upon existing challenges in diabetes mellitus detection and diagnosis
- Assists scholars and students who might like to learn about this area as well as others who may have begun without a formal presentation, with no complex mathematical equations
- Involves exceptional subject coverage and includes the principles needed to understand deep learning

Deep Learning in Diabetes Mellitus Detection and Diagnosis

Edited by

Jyotismita Chaki and Marcin Wozniak

CRC Press
Taylor & Francis Group
Boca Raton London New York

CRC Press is an imprint of the
Taylor & Francis Group, an **informa** business

First edition published 2025
by CRC Press
2385 NW Executive Center Drive, Suite 320, Boca Raton FL 33431

and by CRC Press
4 Park Square, Milton Park, Abingdon, Oxon, OX14 4RN

CRC Press is an imprint of Taylor & Francis Group, LLC

ISBN: 9781032647005 (hbk)
ISBN: 9781032708386 (pbk)
ISBN: 9781032708430 (ebk)

DOI: 10.1201/9781032708430

Typeset in Minion
by codeMantra

Contents

Editor Biographies

Dr. Jyotismita Chaki is an Associate Professor at the School of Computer Science and Engineering, Vellore Institute of Technology, India. Holding a PhD in Engineering from Jadavpur University, Kolkata, her research focuses on Computer Vision, Image Processing, Pattern Recognition, Medical Imaging, Artificial Intelligence, and Machine Learning. With a prolific publication record, Dr. Chaki has authored over 50 international conference and journal papers and contributed to more than ten books as author or editor. In 2024, she was presented among "TOP 2% Scientists in the World". Her editorial roles currently include Editor of Engineering Applications of Artificial Intelligence (Elsevier), Section Editor of PeerJ Computer Science, and Associate Editor for Computer and Electrical Engineering, Array, and Machine Learning with Applications journals (Elsevier). She is a Senior Member of IEEE.

Marcin Wozniak received an MSc degree in applied mathematics, a PhD degree in computational intelligence, a DSc degree in computational intelligence, and Full Professor honours from the President of Poland. He is currently a Full Professor with the Faculty of Applied Mathematics, Silesian University of Technology.

He is a Scientific Supervisor in editions of "The Diamond Grant" and "The Best of the Best" programs for highly talented students from the Polish Ministry of Science and Higher Education. He participated in various scientific projects (as Lead Investigator, Scientific Investigator, Manager, Participant, and Advisor) at Polish, Italian, and Lithuanian universities and projects with applied results in IT industry both funded by the National Centre for Research and Development and abroad. He was a Visiting Researcher with universities in Italy, Sweden, and Germany.

He has authored/coauthored over 300 research papers in international conferences and journals. His current research interests include neural networks with their applications together with various aspects of fuzzy logic and control, applied computational intelligence accelerated by evolutionary computation, and federated learning models.

In 2017, he was awarded by the Polish Ministry of Science and Higher Education with a scholarship for an outstanding young scientist. In 2021 and 2024, he received two awards from the Polish Ministry of Science and Higher Education for research achievements. In 2020, 2021, 2022, and 2023, he was presented among "TOP 2% Scientists in the World" by Stanford University for his career achievements. He is also presented among the Best Computer Science Scientists in Poland by Research.com.

He is an Editorial Board Member or an Editor for Biomedical Signal Processing and Control, Sensors, Machine Learning with Applications, Pattern Analysis, and Applications. He also guest edits special issues, i.e., *IEEE Journal Biomedical and Health Informatics, IEEE Access, Measurement, Sustainable Energy Technologies and Assessments, Frontiers in Human Neuroscience, PeerJ CS, International Journal of Distributed Sensor Networks, Computational Intelligence and Neuroscience, Journal of Universal Computer Science*, etc. He is a Session Chair at various international conferences and symposiums, including *IEEE Symposium Series on Computational Intelligence, IEEE Congress on Evolutionary Computation*, and *International Joint Conference on Neural Networks.*

List of Contributors

Chukwuma Chinaza Adaobi
School of Public Health and Allied Science
Catholic University of Ghana
Sunyani, Ghana

Elaheh Afsaneh
Independent Researcher
Iran

N. P. G. Bhavani
ECE Department
Saveetha Institute of Medical and Technical Sciences
Chennai, India

N. G. Bhuvaneswari Amma
School of Computer Science and Engineering
Vellore Institute of Technology
Vellore, India

Akshat Bokdia
School of Computer Science and Engineering
Vellore Institute of Technology
Vellore, India

SuQun Cao
Division of Science and Technology
Huaiyin Institute of Technology
Huaiyin, People's Republic of China

Ravindra Changala
Department of Computer Science and Engineering
Guru Nanak Institutions Technical Campus
Telangana, India

Mohammad Dehghani
Electrical and Computer Engineering Department
University of Tehran
Tehran, Iran

Lipika Dinda
Department of Computer Science and Engineering
National Institute of Technology
Jamshedpur, India

Mohadeseh Zarei Ghobadi
Independent Researcher
Iran

Dibyajyoti Ghosh
VIT Business School
Vellore Institute of Technology
Vellore, India

C. Gunavathi
School of Computer Science and Engineering
Vellore Institute of Technology
Vellore, India

N. Janaki
EEE Department
Vels Institute of Science, Technology and Advanced Studies
Chennai, India

S. Kanakaprabha
Department of Computer Science and Engineering
Rathinam Technical Campus
Coimbatore, India

D. Kavitha
Department of Information Technology
PVP Siddhartha Institute of Technology
Vijayawada, India

M. Karthikeyan
Department of Electrical and Electronics Engineering
SRM Institute of Science and Technology
Ramapuram, India

Prajwal Kulkarni
School of Computer Science and Engineering
Vellore Institute of Technology
Vellore, India

Siri R. Kulakarni
School of Computer Science and Engineering
Vellore Institute of Technology
Vellore, India

G. Ganesh Kumar
Department of Information Technology
Hindusthan Institute of Technology
Valley Campus, India

Lella Kranthi Kumar
School of Computer Science and Engineering
VIT-AP University
Amaravati, India

Kezhi Li
Institute of Health Informatics
University College London
London, United Kingdom

S. V. Suresh Babu Matla
Department of Computer Science and Engineering
Koneru Lakshmaiah Educational Foundation
Vijayawada, India

Atianashie A. Miracle
School of Research and Graduate Studies
Catholic University of Ghana
Sunyani, Ghana

Ayush Negi
School of Computer Science and Engineering
Vellore Institute of Technology
Vellore, India

Govind Narayan Patel
Department of Computer Science and Engineering
National Institute of Technology
Jamshedpur, India

Chengzhe Piao
Institute of Health Informatics
University College London
London, United Kingdom

R. S. Ponmagal
CSE Department
SRM Institute of Science and Technology
Kattankulathur, India

Jitesh Pradhan
Department of Computer Science and Engineering
National Institute of Technology
Jamshedpur, India

B. Ramachandra Reddy
Dept. of Computer Science and Engineering
National Institute of Technology
Jamshedpur, India

R. Rengaraj
Department of Electrical and Electronics Engineering
Sri Sivasubramaniya Nadar College of Engineering
Kalavakkam, India

Muskan Sah
School of Computer Science and Engineering
Vellore Institute of Technology
Vellore, India

Pranav Sehgal
School of Computer Science and Engineering
Vellore Institute of Technology
Vellore, India

Shrish Srinivasan
Department of Education
Kamala Niketan Montessori School
Trichy, India

K. Sujatha
EEE Department
Dr. MGR Educational and Research Institute
Chennai, India

Akila Victor
School of Computer Science and Engineering
Vellore Institute of Technology
Vellore, India

Katarzyna Wiltos
Faculty of Applied Mathematics
Silesian University of Technology
Gliwice, Poland

CHAPTER 1

Introduction to Diabetes Mellitus Detection and Diagnosis Using Deep Learning

Jyotismita Chaki and Dibyajyoti Ghosh

1.1 THE GLOBAL BURDEN OF DIABETES MELLITUS

Diabetes mellitus (DM), often simply referred to as diabetes, is a chronic metabolic disorder characterized by abnormally high blood sugar (glucose) levels. This high blood sugar arises due to defects in either the production of insulin by the pancreas, the body's response to insulin (insulin resistance), or a combination of both.

Here are the key aspects of DM [1]:

- **Chronic:** Diabetes is a long-term condition that requires ongoing management throughout a person's life.
- **Metabolic Disorder:** It disrupts the body's normal metabolic processes, particularly how it uses glucose for energy.
- **Hyperglycemia:** The hallmark feature of diabetes is elevated blood sugar levels. Normally, the body tightly regulates blood sugar within

DOI: 10.1201/9781032708430-1

a narrow range. In diabetes, this regulation is impaired, leading to chronic hyperglycemia.

- **Defects in Insulin:** Insulin is a hormone produced by the pancreas that acts like a key, allowing glucose to enter cells from the bloodstream and be used for energy. In diabetes, either the pancreas doesn't produce enough insulin, the body's cells become resistant to insulin's effects, or both.

There are different types of diabetes with varying causes, but all share the common thread of hyperglycemia. Early detection and proper management are crucial to prevent serious health complications associated with chronic high blood sugar.

1.2 UNVEILING THE DIFFERENT FACES OF DIABETES MELLITUS (DM)

While the core issue of high blood sugar defines diabetes, the underlying causes can differ. The most common types of DM are as follows [2]:

1.2.1 Type 1 Diabetes (T1D)

The body's immune system mistakenly attacks the insulin-producing beta cells in the pancreas, leading to insulin deficiency. This is typically diagnosed in childhood or young adulthood and requires lifelong insulin injections or an insulin pump to regulate blood sugar levels.

1.2.2 Type 2 Diabetes (T2D)

The body's cells become resistant to the action of insulin, leading to a buildup of glucose in the blood. The pancreas may still produce insulin, but it becomes insufficient. This can develop at any age, but the risk increases with age, family history, and lifestyle factors like obesity and physical inactivity and may involve lifestyle modifications like diet, exercise, and weight control. Medications like oral medications or injectable insulin may also be necessary.

1.2.3 Gestational Diabetes (GDM)

This temporary form of diabetes develops during pregnancy due to hormonal changes that can affect insulin sensitivity. This is typically diagnosed during the second or third trimester of pregnancy and is often controlled through dietary adjustments and exercise. In some cases, medication may be needed. GDM usually resolves after delivery, but women with GDM have a higher risk of developing type 2 diabetes later in life.

1.2.4 Other Specific Types of DM

Monogenic diabetes (MODY) is a group of genetically inherited forms of diabetes caused by mutations in specific genes. Maturity-onset diabetes of the young (MODY) is a subtype of MODY that typically starts before age 25. Neonatal diabetes is a rare form of diabetes that manifests in the first few months of life. Secondary diabetes is caused by diseases or conditions that damage the pancreas or affect insulin action (e.g., cystic fibrosis, pancreatitis, and medications).

1.3 DIABETES: A LOOMING GLOBAL CRISIS

DM is not just a health concern; it's a growing global threat. The prevalence of diabetes is exploding worldwide, posing a significant burden on individuals, healthcare systems, and economies. Let us delve into the alarming statistics reported by the International Diabetes Federation (IDF), the leading global organization for diabetes [3].

- **Sheer Numbers:** According to the IDF Diabetes Atlas (2021 edition), an estimated **537 million adults** (aged 20–79 years) are currently living with diabetes globally. That's approximately **one in ten adults** on the planet.
- **Projections Paint a Grim Picture:** The IDF predicts a staggering rise in diabetes cases, with the number expected to reach **643 million by 2030** and a shocking **783 million by 2045**. This translates to a **46% increase** in just over two decades.
- **Undiagnosed Cases Fuel the Fire:** A significant concern is the high number of undiagnosed cases. The IDF estimates that **almost half (44.7%)** of adults with diabetes are unaware that they have the condition. This lack of awareness delays proper management and increases the risk of complications.
- **Unequal Burden:** The burden of diabetes is not evenly distributed. **Over three in four adults** with diabetes live in **low- and middle-income countries**. This disparity highlights the need for improved access to healthcare and diabetes education in these regions.

These figures paint a clear picture: diabetes is a global health crisis with a rapidly rising trajectory. Immediate and concerted efforts are needed to prevent new cases, ensure early diagnosis, and provide effective management to those living with diabetes.

1.4 A LOOK AT ASSOCIATED HEALTH COMPLICATIONS BY DIABETES

While chronic high blood sugar is the hallmark of diabetes, the true danger lies in the cascade of health complications it can trigger over time. Left unmanaged, diabetes can lead to a multitude of serious health issues, significantly impacting a person's quality of life. Here's a glimpse into some of the most concerning complications [4]:

1.4.1 Microvascular Complications: These Affect Small Blood Vessels throughout the Body

- **Diabetic retinopathy:** Damage to the blood vessels in the retina, the light-sensitive layer of the eye, can lead to vision problems, including blindness.
- **Diabetic nephropathy:** High blood sugar damages the delicate filters in the kidneys, leading to chronic kidney disease and potentially kidney failure.
- **Diabetic neuropathy:** Nerve damage caused by high blood sugar can affect sensation, leading to numbness, pain, tingling, and burning in the hands and feet. This can also affect digestion and sexual function.

1.4.2 Macrovascular Complications: These Involve Damage to Large Blood Vessels

- **Cardiovascular disease (CVD):** Diabetes significantly increases the risk of heart attack, stroke, and peripheral artery disease (PAD) which affects blood flow to the legs.
- **Peripheral arterial disease (PAD):** High blood sugar and neuropathy can lead to poor circulation in the legs, increasing the risk of infections and non-healing wounds, potentially leading to amputations.

1.4.3 Other Complications

Diabetes can also contribute to:

- **Hearing problems:** High blood sugar can damage nerves related to hearing.
- **Skin problems:** Diabetic neuropathy can decrease sweating, increase susceptibility to infections, and slow wound healing.
- **Mental health issues:** Depression and anxiety are more common in people with diabetes due to the chronic nature of the disease and the challenges of managing it.

Early diagnosis and effective management of diabetes are crucial to prevent or delay the onset of these complications. Maintaining healthy blood sugar levels through lifestyle modifications, medication, and regular monitoring is essential for long-term well-being.

1.5 TRADITIONAL METHODS OF DIABETES DETECTION AND DIAGNOSIS: A GLIMPSE INTO THE PAST

For decades, healthcare professionals have relied on specific tools to diagnose and monitor diabetes. While these methods have played a vital role in managing the disease, they are not without limitations. Let's delve into the traditional approaches used for diabetes detection and diagnosis [5,6].

1.5.1 Fasting Blood Glucose (FBG) and Oral Glucose Tolerance Test (OGTT)

- **FBG test:** This test measures blood sugar levels after a person hasn't eaten or drunk anything (except water) for at least 8 hours, typically done first thing in the morning. A fasting blood sugar level above a certain threshold (usually 126 mg/dL or 7.0 mmol/L) can indicate diabetes [7].
- **OGTT:** This 2-hour test measures blood sugar levels after drinking a sugary beverage. Blood sugar is checked before the drink (fasting) and then again at 1 and 2 hours after consumption. Elevated levels at specific points during the test suggest impaired glucose tolerance or diabetes [7].

While FBG and OGTT are considered the gold standard, they have several drawbacks:

- **Inconvenience:** Both tests require fasting, which can be disruptive to daily routines. OGTT, especially, involves multiple blood draws and waiting periods, making it time-consuming.
- **Cost:** These tests can be expensive, depending on insurance coverage.
- **Potential inaccuracy:** Stress, medications, and recent meals can affect the accuracy of blood sugar levels, leading to potential misdiagnosis.
- **Limited scope:** These tests only provide a snapshot of blood sugar levels at a specific point in time. They don't capture the full picture of how well the body manages blood sugar over an extended period.

1.5.2 Glycated Hemoglobin (HbA1c) Test: A Monitoring Tool

The HbA1c test measures the percentage of hemoglobin (a protein in red blood cells) that has attached to glucose. Since red blood cells live for about 3 months, HbA1c reflects average blood sugar control over the previous 2–3 months. This makes it a valuable tool for monitoring diabetes control and treatment effectiveness [8].

1.5.3 Other Diagnostic Tools

- **Urine tests:** These may be used to check for the presence of ketones, a byproduct produced by the body when it breaks down fat for energy due to a lack of insulin. Elevated ketones can signal a serious complication called diabetic ketoacidosis.
- **Physical examination:** A doctor may look for physical signs of diabetes, such as excessive thirst, frequent urination, unexplained weight loss, and slow-healing wounds.

While traditional methods have served their purpose, the limitations are evident. The need for **improved, non-invasive, and cost-effective** methods for early detection and diagnosis of diabetes is critical. This is where advancements in deep learning offer a promising future. With the ability to analyze vast datasets and identify subtle patterns, deep learning holds immense potential to revolutionize how we diagnose and manage diabetes.

1.6 DEEP LEARNING: USHERING IN A NEW ERA OF MEDICAL DIAGNOSIS

The field of medicine is on the cusp of a transformative revolution, driven by the power of deep learning (DL). This subfield of artificial intelligence

(AI) is rapidly changing the landscape of disease diagnosis and management. Let's embark on a journey to understand how deep learning works and its immense potential in the realm of diabetes detection [9].

Deep learning draws inspiration from the intricate structure and function of the human brain. At its core lies the concept of artificial neural networks (ANNs). These are computational models that mimic the interconnected network of neurons in the brain.

ANNs consist of interconnected layers of artificial neurons. Each neuron receives input from other neurons, performs a simple mathematical operation, and transmits a processed output signal. By stacking these layers upon each other, we create deep neural networks, capable of learning complex patterns from vast amounts of data.

1.6.1 Deep Learning Architectures: Specialized Tools for Diverse Tasks

Deep learning encompasses a variety of architectures, each tailored for specific tasks:

- **Convolutional neural networks (CNNs):** These excel at image recognition. CNNs are adept at extracting features from images, making them ideal for analyzing medical scans like X-rays, CT scans, and retinal images [10].
- **Recurrent neural networks (RNNs):** These are designed to handle sequential data, where the order of information matters. RNNs are particularly useful for analyzing time series data like gene sequences or electronic health records [11].

1.6.2 Deep Learning in Healthcare: A Beacon of Hope

The ability of deep learning models to learn intricate patterns from massive datasets holds immense potential for medical diagnosis. Here's how [12]:

- **Enhanced accuracy:** Deep learning models can analyze vast amounts of medical data, including images, lab results, and patient history, with a level of detail and accuracy that surpasses human capabilities. This can lead to earlier and more precise diagnoses.
- **Improved efficiency:** Deep learning can automate tasks like image analysis and data interpretation, freeing up healthcare professionals to focus on patient care.

- **Personalized medicine:** Deep learning models can leverage large datasets to identify subtle patterns and risk factors, paving the way for personalized medicine approaches tailored to individual patients.

1.6.3 Ethical Considerations: Navigating the Uncharted Territory

While the potential of deep learning in healthcare is undeniable, ethical considerations need careful attention [13]:

- **Bias:** Deep learning models are only as good as the data they are trained on. Biases present in the training data can lead to biased outputs, potentially exacerbating existing healthcare disparities.
- **Transparency and explainability:** Deep learning models can be complex "black boxes". Ensuring transparency in how models arrive at their predictions is crucial for building trust and addressing potential bias.
- **Regulation and oversight:** As deep learning models become integrated into clinical workflows, robust regulatory frameworks are necessary to ensure patient safety and data privacy.

The journey with deep learning in healthcare has just begun. By acknowledging the challenges and working toward responsible development and implementation, we can harness the power of this technology to revolutionize the diagnosis and management of diabetes and other medical conditions.

1.7 DEEP LEARNING FOR DIABETES DETECTION AND DIAGNOSIS

1.7.1 Deep Learning with Medical Images: Unveiling Diabetic Retinopathy through Retinal Fundus Analysis

Diabetic retinopathy (DR) is a leading cause of vision loss in adults with diabetes. Early detection and treatment are critical to prevent vision impairment. Deep learning, particularly convolutional neural networks (CNNs), offers a promising approach for analyzing retinal fundus images to detect DR effectively [14,15].

1.7.1.1 Demystifying DR Detection with CNNs

Here's a deeper dive into how CNNs can be utilized for detecting DR in retinal fundus images:

a. **Data preprocessing:** Retinal fundus images often exhibit variations in brightness, contrast, and field of view. Preprocessing steps ensure consistency and improve model performance. This might involve:

- **Resizing:** Images are resized to a standard size for efficient processing by the CNN.
- **Normalization:** Brightness and contrast are adjusted to enhance relevant features and reduce noise.
- **Macula centering:** The region of interest, the macula (central part of the retina), is automatically identified and centered within the image.

b. **Feature extraction:** The core strength of CNNs lies in their ability to automatically extract features from images. In DR detection, these features are crucial for identifying early signs of the disease:

- **Blood vessel segmentation:** CNNs can segment the intricate network of blood vessels in the retina. Abnormalities in blood vessel width, caliber, and leakage can indicate DR.
- **Microaneurysm detection:** Early signs of DR include the appearance of microaneurysms, tiny bulges in retinal blood vessels. CNNs can effectively detect these subtle changes.
- **Hemorrhage and exudate identification:** Bleeding (hemorrhages) and waxy white patches (exudates) within the retina are signs of advanced DR. CNNs are adept at identifying these features.

c. **Classification:** Based on the extracted features, the CNN classifies the image into one of several categories:

- **Normal:** The retina shows no signs of DR.
- **Mild DR:** Early signs of DR are present, such as microaneurysms.
- **Moderate DR:** More advanced signs of DR are evident, including hemorrhages and exudates.
- **Severe DR:** The presence of significant damage to the retina due to DR.

1.7.1.2 Advantages of CNNs for DR Detection

- **Automated feature extraction:** Unlike traditional image analysis methods that require manual feature engineering, CNNs automatically learn the most relevant features for DR detection. This eliminates human bias and streamlines the process.
- **Scalability:** CNNs excel at handling large datasets of retinal fundus images. This is crucial for training robust models that can generalize well to unseen data.
- **Improved accuracy:** Studies have shown that CNN-based models can achieve high accuracy in DR detection, rivaling or surpassing the performance of human experts, especially for detecting early signs of the disease.

1.7.1.3 How CNNs "See" Diabetic Retinopathy

Imagine a CNN as a multi-layered hierarchy. The initial layers focus on detecting basic edges and shapes in the image. As we move toward deeper layers, the network combines these lower-level features to form more complex representations. In the context of DR detection, these higher-level features might represent specific patterns associated with blood vessel abnormalities, microaneurysms, or hemorrhages. Finally, the final layer of the CNN integrates all the learned features to classify the image as normal or indicative of DR.

Deep learning with CNNs offers a powerful tool for analyzing retinal fundus images and detecting DR. By automating feature extraction and achieving high accuracy, CNNs hold immense promise for early DR detection, enabling timely intervention and improved patient outcomes. As research continues to explore advanced CNN architectures and larger datasets, this technology has the potential to revolutionize the diagnosis and management of DR.

1.7.2 Deep Learning with Electronic Health Records (EHRs): Mining a Treasure Trove for Diabetes Risk Prediction

Electronic health records (EHRs) have revolutionized healthcare by providing a centralized platform for storing a wealth of patient information. This rich data source, encompassing demographics, lab test results, medication history, diagnoses, and clinical notes, holds immense potential for early disease detection and risk prediction. Deep learning algorithms, particularly recurrent neural networks (RNNs) and long short-term memory (LSTM) networks, can unlock the secrets hidden within EHR data to predict the risk of developing diabetes [16,17].

1.7.2.1 EHRs: A Goldmine of Patient Information

EHRs are digital records that contain a comprehensive history of a patient's medical encounters. These data include:

- **Demographics:** Age, gender, ethnicity, and family history.
- **Lab test results:** Blood sugar levels, cholesterol levels, and HbA1c levels.
- **Medications:** Prescribed medications and their dosages.
- **Diagnoses:** Documented medical conditions, including previous diagnoses related to diabetes risk factors.
- **Clinical notes:** Physician observations, progress notes, and treatment plans.

1.7.2.2 RNNs and LSTMs: Decoding the Sequence in EHR Data

RNNs are a type of deep learning architecture designed to handle sequential data, where the order of information is crucial. This characteristic makes them ideal for analyzing EHR data, which is inherently sequential, with each patient encounter adding to the overall medical history.

However, traditional RNNs struggle with learning long-term dependencies within long sequences. Here's where LSTM networks come into play. LSTMs are a specific type of RNN architecture equipped with internal mechanisms to remember and utilize information from earlier parts of the sequence. This allows LSTMs to capture the temporal relationships within EHR data and identify patterns that may indicate an increased risk of developing diabetes.

1.7.2.3 Harnessing Deep Learning for Diabetes Risk Prediction

Deep learning models trained on EHR data can learn complex patterns and relationships between various factors that contribute to diabetes risk. Here's how:

- **Data preprocessing:** EHR data needs to be cleaned, formatted, and transformed into a suitable format for the deep learning model. This may involve handling missing values, converting categorical data into numerical representations, and defining time intervals for sequential analysis.
- **Feature engineering:** While LSTMs can learn features to some extent, some studies may involve incorporating domain knowledge by extracting additional features like body mass index (BMI) from EHR data.
- **Model training:** The deep learning model, typically an LSTM network, is trained on a large dataset of EHRs from patients with and without diabetes. During training, the model learns to identify patterns in the data that differentiate individuals at higher risk of developing diabetes.
- **Risk prediction:** Once trained, the model can be used to predict the risk of developing diabetes for a new patient. The model analyzes the patient's EHR data and outputs a score or probability indicating their risk level.

1.7.2.4 Benefits of Deep Learning for Diabetes Risk Prediction

- **Early identification:** Deep learning models can identify individuals at high risk of developing diabetes even before they exhibit symptoms. This allows for early intervention and preventive measures to be implemented.
- **Personalized risk assessment:** Deep learning models can consider a broader range of variables compared to traditional risk assessment tools, leading to more personalized risk assessments.
- **Scalability:** Deep learning models can efficiently handle large and complex datasets of EHRs, making them suitable for large-scale risk prediction efforts.

1.7.2.5 Challenges and Considerations

Despite the promising potential, there are challenges associated with using deep learning for diabetes risk prediction:

- **Data quality:** The accuracy of the model heavily relies on the quality and completeness of data within EHRs.
- **Data privacy:** Ensuring patient data privacy and security is paramount when using EHR data for deep learning models.
- **Model explainability:** Understanding how the model arrives at its risk predictions is important for building trust in the technology and mitigating potential biases.

Deep learning offers a powerful tool for analyzing EHR data and predicting the risk of developing diabetes. By leveraging this technology, healthcare professionals can proactively identify high-risk individuals and implement targeted preventive measures. As research continues to address challenges and optimize models, deep learning has the potential to revolutionize diabetes risk prediction and improve overall population health outcomes.

1.7.3 Deep Learning and Wearable Devices: Ushering in a Personalized Era of Diabetes Management

The landscape of health monitoring is undergoing a significant transformation with the rise of wearable devices. These compact, user-friendly devices, including continuous glucose monitors (CGMs), fitness trackers, and smartwatches, are empowering individuals to track various health metrics in real time. Deep learning, with its ability to analyze vast amounts of data, presents a unique opportunity to leverage wearable device data for effective diabetes management [18,19].

1.7.3.1 The Wearable Tech Revolution

A growing number of people are embracing wearable devices for health monitoring. Here are some prominent examples:

- **Continuous glucose monitors (CGMs):** These minimally invasive devices measure blood sugar levels in the interstitial fluid (fluid between cells) throughout the day, providing a more comprehensive picture of glucose fluctuations compared to traditional finger pricking.
- **Fitness trackers and smartwatches:** These devices track various health parameters like heart rate, sleep patterns, and activity levels.

While not directly measuring blood sugar, these data can offer valuable insights into factors influencing glucose levels.

1.7.3.2 Deep Learning for Blood Sugar Tracking and Trend Prediction

Deep learning models can analyze data from wearable devices to support diabetes management in several ways:

- **Blood sugar level tracking with CGMs:** Deep learning models can analyze CGM data to identify patterns and trends in blood sugar levels. This can help individuals with diabetes understand how their body reacts to food, exercise, and other factors.
- **Predicting glucose trends:** By analyzing historical CGM data and incorporating user-specific information (diet, medication), deep learning models can predict future blood sugar trends. This allows for proactive adjustments to insulin dosages or dietary choices.
- **Personalized insights:** Deep learning models can personalize insights by considering individual variations in physiology and lifestyle habits. This personalized approach can lead to more effective and targeted diabetes management strategies.

1.7.3.3 Challenges and Considerations

While the potential of deep learning with wearable devices is undeniable, here are some challenges to address:

- **Data quality:** The accuracy of deep learning models hinges on the quality and consistency of data collected from wearable devices. Factors like sensor calibration and user compliance can impact data quality.
- **Data security and privacy:** Safeguarding sensitive health data collected by wearables is paramount. Robust security measures and user privacy protocols are essential.
- **Model explainability:** Understanding how deep learning models arrive at their predictions is crucial for building trust and ensuring the models don't perpetuate biases based on historical data.

1.7.3.4 Combining Wearable Data with EHRs: A Holistic View

The true power of deep learning lies in its ability to integrate data from various sources. Combining wearable device data with EHRs offers a more holistic perspective on an individual's health:

- **Richer context:** EHRs contain valuable information like lab test results, medication history, and diagnoses. Integrating these data with wearable device data allows for a more comprehensive understanding of factors influencing blood sugar control.
- **Improved prediction accuracy:** By combining diverse data sources, deep learning models can potentially achieve higher accuracy in predicting blood sugar trends and identifying potential complications.
- **Tailored interventions:** Access to a broader range of data can enable healthcare professionals to develop more personalized and effective treatment plans for individuals with diabetes.

1.7.3.5 The Future of Wearables and Deep Learning

Deep learning with wearable device data has the potential to revolutionize diabetes management. By offering real-time insights, predicting trends, and enabling personalized interventions, this technology empowers individuals with diabetes to take a more proactive role in managing their health. As research progresses in data security, model explainability, and integration with EHRs, deep learning and wearable devices hold immense promise for creating a future of personalized and data-driven diabetes care.

1.8 CHALLENGES ON THE ROAD TO DEEP LEARNING SOLUTIONS FOR DIABETES DETECTION AND DIAGNOSIS

Deep learning offers a powerful toolkit for analyzing medical data and revolutionizing diabetes detection and diagnosis. However, the path to widespread adoption is paved with significant challenges. Here, we delve into four key hurdles that need to be addressed [20,21]:

1.8.1 Data Availability and Quality—The Foundation of Deep Learning

Deep learning models are data-driven. Their success hinges on the quality and quantity of data used for training. In the context of diabetes detection and diagnosis, this translates to several specific challenges:

- **Limited dataset size:** Large, diverse datasets of medical images (retinal fundus images), EHRs, and wearable device data are essential for training robust deep learning models. However, collecting such datasets can be expensive and time-consuming. Additionally, patient privacy concerns and institutional data silos can further restrict access to valuable data.
- **Data bias:** Deep learning models are susceptible to perpetuating biases present in the training data. If datasets predominantly represent a specific demographic or disease severity, the model might not perform well on patients with different characteristics. This can exacerbate existing healthcare disparities.
- **Data quality issues:** Medical data can be riddled with inconsistencies, missing values, and errors. Deep learning models are sensitive to such issues, and data cleaning and preprocessing can be a complex and resource-intensive task.

1.8.1.1 Addressing the Data Challenge

- **Data sharing initiatives:** Fostering collaboration between healthcare institutions, research groups, and technology companies can facilitate the creation of larger and more diverse datasets.
- **Synthetic data generation:** Techniques like generative adversarial networks (GANs) can be employed to create synthetic medical data that augments existing datasets and mitigates privacy concerns.
- **Data standardization:** Establishing standardized data collection and annotation protocols across healthcare institutions can improve data quality and facilitate sharing.

1.8.2 Model Explainability and Interpretability—Demystifying the Black Box

Deep learning models are often referred to as "black boxes" due to the complexity of their internal workings. This lack of transparency poses several challenges:

- **Limited trust:** Healthcare professionals need to understand how deep learning models arrive at their predictions to build trust in the technology and ensure it aligns with their clinical judgment.

- **Debugging and bias detection:** Without interpretability, it's difficult to identify and address potential biases embedded within the model.
- **Regulatory hurdles:** Regulatory bodies might hesitate to approve deep learning models for clinical use if their decision-making processes are not explainable.

1.8.2.1 Enhancing Model Explainability

- **Explainable AI (XAI) techniques:** Research in XAI aims to develop methods for interpreting deep learning models. This can involve techniques like feature attribution methods that highlight which aspects of the data played a role in the model's prediction.
- **Visualization tools:** Developing visualizations that translate the model's internal workings into human-understandable representations can aid in understanding its decision-making process.
- **Human-in-the-loop approach:** Deep learning models can be integrated into a human-in-the-loop system where healthcare professionals review and potentially adjust the model's recommendations, fostering trust and transparency.

1.8.3 Integration with Clinical Workflow—Bridging the Gap between Research and Practice

For deep learning models to have a real-world impact on diabetes detection and diagnosis, seamless integration into existing clinical workflows is crucial. Here are some key considerations:

- **Workflow disruption:** Introducing new technologies can disrupt established workflows. Deep learning models need to be designed and implemented in a way that complements, rather than hinders, the work of healthcare professionals.
- **User interface and usability:** The user interface for interacting with deep learning models in a clinical setting needs to be user-friendly and intuitive for healthcare professionals with varying levels of technical expertise.
- **Infrastructure and technical support:** Hospitals and clinics might require upgrades to their IT infrastructure to support the implementation and integration of deep learning models.

1.8.3.1 Ensuring Smooth Integration

- **Co-design with clinicians:** Deep learning models should be developed in collaboration with clinicians to ensure they address real-world needs and seamlessly integrate into existing workflows.
- **User-centered design:** The user interface needs to be designed with the specific needs and workflow of healthcare professionals in mind.
- **Investment in infrastructure:** Healthcare institutions need to consider investments in IT infrastructure and technical support to enable the adoption of deep learning models.

1.9 FUTURE DIRECTIONS FOR DEEP LEARNING IN DIABETES CARE

Deep learning has emerged as a transformative force in healthcare, with immense potential to revolutionize diabetes care. While challenges persist, ongoing research and development efforts are paving the way for exciting future directions [22]. Here, we explore some key areas where deep learning is poised to make a significant impact:

1.9.1 Enhanced Disease Detection and Risk Prediction

- **Multimodal deep learning:** Deep learning models can be trained on a combination of data sources, such as retinal images, EHR data, and wearable device data. This multimodal approach can lead to more accurate and comprehensive risk prediction for diabetes and its complications.
- **Early detection of complications:** Deep learning models can analyze medical images and sensor data to identify early signs of diabetic complications like retinopathy, nephropathy, and neuropathy. This allows for early intervention and potentially prevents complications from progressing.
- **Personalized risk stratification:** Deep learning models can consider a broader range of individual factors to create personalized risk profiles for diabetes development and complications. This enables more targeted preventive measures and treatment strategies.

1.9.2 Personalized Management and Treatment Optimization

- **Predictive blood sugar control:** Deep learning models can analyze real-time and historical data from CGMs and other wearables to predict future blood sugar trends. This empowers individuals with diabetes to make informed decisions about diet, exercise, and medication adjustments.
- **Closed-loop insulin delivery systems:** Deep learning algorithms can be integrated into closed-loop insulin delivery systems (artificial pancreas) to continuously monitor blood sugar levels and automatically adjust insulin delivery in real time. This can lead to tighter blood sugar control and improved quality of life.
- **Virtual diabetes management assistants:** Deep learning-powered virtual assistants can provide personalized guidance, support, and education to individuals with diabetes, promoting self-management and adherence to treatment plans.

1.9.3 Deep Learning for Clinical Decision Support and Research

- **Clinical decision support systems (CDSS):** Deep learning models can be integrated into CDSS to offer real-time insights and recommendations to healthcare professionals at the point of care. This can support informed decision-making and improve patient outcomes.
- **Drug discovery and development:** Deep learning can be utilized for tasks like analyzing vast datasets of patient data and molecular structures to accelerate drug discovery and development for diabetes and its complications.
- **Real-world data analysis:** Deep learning can analyze real-world data from electronic health records and wearables to gain deeper insights into disease progression, treatment effectiveness, and patient outcomes in real-world settings.

1.9.4 Ethical Considerations and Responsible Development

- **Addressing bias and fairness:** Mitigating bias in deep learning models for diabetes care is crucial to ensure equitable access to these technologies and prevent the exacerbation of healthcare disparities.

Techniques like fairness-aware model training and diverse data collection are essential.

- **Transparency and explainability:** Enhancing model explainability will build trust and ensure healthcare professionals understand how deep learning models arrive at their predictions. This is critical for integrating these models seamlessly into clinical workflows.
- **Data privacy and security:** Implementing robust data security measures and ensuring compliance with data privacy regulations is paramount when using deep learning models in healthcare.

1.9.5 Explainable AI (XAI): Building Trust and Understanding

One of the major challenges associated with deep learning models is their inherent "black box" nature. The complex internal workings of these models can be opaque, making it difficult to understand how they arrive at their predictions. In the context of diabetes care, where crucial decisions are based on model outputs, a lack of transparency can hinder trust and adoption.

1.9.5.1 The Need for Explainability

- **Building trust:** Healthcare professionals need to understand the rationale behind a model's prediction to trust its recommendations and integrate it effectively into their clinical workflow.
- **Debugging and bias detection:** Without interpretability, it's challenging to identify and address potential biases within the model. This is critical to ensure fairness and prevent the exacerbation of healthcare disparities.
- **Regulatory approval:** Regulatory bodies might hesitate to approve deep learning models for clinical use if their decision-making processes are not explainable.

1.9.5.2 XAI Techniques for Deep Learning in Diabetes Care

- **Feature attribution methods:** These techniques highlight which features in the data (e.g., specific patterns in retinal images) contributed most significantly to the model's prediction. This can provide insights into the model's reasoning.

- **Visualization tools:** Developing visualizations that translate the model's internal workings into human-understandable representations can aid in understanding its decision-making process. For example, visualizing which parts of a retinal image led the model to predict DR.
- **Model-agnostic explanations:** These techniques can be applied to any deep learning model, regardless of its architecture, to provide explanations for its predictions.

1.9.6 Federated Learning: Empowering Collaboration with Data Privacy

Deep learning models are data-driven, and access to large, diverse datasets is crucial for training robust models. However, patient privacy concerns often pose a barrier to data sharing. Federated learning offers a potential solution to this challenge.

1.9.6.1 The Power of Federated Learning

- **Privacy-preserving collaboration:** Federated learning allows multiple institutions to collaborate on training a deep learning model without directly sharing patient data. Each institution trains the model locally on its own data and then shares only the model updates (weights) with a central server. This allows the model to learn from the combined knowledge of all participating institutions while protecting patient privacy.
- **Enhanced generalizability:** By leveraging data from diverse sources, federated learning can lead to models that are more generalizable and perform well on unseen data. This is crucial for developing models that can be effectively applied in real-world settings with heterogeneous patient populations.
- **Accelerated development:** Federated learning enables parallel training on multiple devices or servers, potentially speeding up the development and deployment of deep learning models for diabetes care.

1.9.6.2 Challenges and Considerations for Federated Learning

- **Technical challenges:** Developing robust and efficient federated learning algorithms requires addressing challenges like communication overhead and non-IID (non-independent and identically distributed) data, where data distributions across institutions might vary.

- **Standardization and governance:** Standardized protocols for data formatting, model updates, and privacy protection are necessary for successful implementation of federated learning in healthcare.

1.9.6.3 The Road Ahead: A Collaborative Effort

The future of deep learning in diabetes care hinges on a collaborative effort between researchers, clinicians, technology developers, and regulatory bodies. By addressing the existing challenges, fostering responsible development practices, and focusing on ethical considerations, deep learning has the potential to transform diabetes care by:

- **Empowering individuals with diabetes:** To take a more active role in managing their health.
- **Enabling healthcare professionals:** To make more informed decisions and improve patient outcomes.
- **Advancing research and development:** In diabetes prevention, treatment, and cure discovery.

As we navigate the exciting future of deep learning, we must prioritize patient well-being, ensure equitable access to these technologies, and harness the power of deep learning to create a future where diabetes can be effectively managed and potentially even prevented.

1.10 CONCLUSIONS

This chapter discusses the introduction to DM detection and diagnosis using DL. Deep learning is rapidly transforming the landscape of diabetes detection and diagnosis. By leveraging its ability to analyze vast amounts of medical data, deep learning offers a powerful toolkit for (i) earlier and more accurate detection: Deep learning models can analyze retinal fundus images, EHR data, and wearable device data to identify early signs of diabetes and its complications, leading to earlier intervention and improved outcomes; (ii) personalized risk stratification: Deep learning models can consider a broader range of individual factors to create personalized risk profiles for developing diabetes and its complications. This enables more targeted preventive measures and treatment strategies; (iii) improved clinical decision support: Deep learning models can be integrated into clinical decision support systems, offering real-time insights and recommendations to healthcare professionals at the point of care. However, challenges remain: (i) data availability and quality: Access to large, diverse, and well-annotated datasets is crucial for training robust models. Addressing data privacy concerns and ensuring data quality are essential steps; (ii) model explainability and interpretability: Deep learning models can be opaque, hindering trust and hindering the identification of potential biases. Research in explainable AI (XAI) is crucial for building trust and ensuring fairness; (iii) integration with clinical workflow: Seamless integration of deep learning models into existing clinical workflows is necessary for real-world adoption. Collaboration with clinicians throughout the development process is key; (iv) regulatory hurdles: Ensuring adherence to data privacy regulations and obtaining regulatory approvals for clinical use require careful consideration. The future of deep learning in diabetes care hinges on a collaborative effort between researchers, clinicians, technology developers, and regulatory bodies. By (i) developing explainable AI techniques: Building trust and understanding through transparency in model decision-making; (ii) exploring privacy-preserving techniques: Utilizing federated learning to enable collaboration on data analysis while maintaining patient privacy; (iii) prioritizing ethical considerations: Mitigating bias and ensuring equitable access to these technologies. Deep learning holds immense promise for revolutionizing diabetes care by empowering individuals, enabling informed clinical decisions, and paving the way for a future with better prevention, management, and potentially even a cure for diabetes.

REFERENCES

1. *Diabetes* (https://www.who.int/health-topics/diabetes?gad_source=1&gclid=CjwKCAjwrcKxBhBMEiwAIVF8rBs4rkBL_c_qLQKk97AcE8T8ozYaph-wamznfMHjFLt63Qhmn1DQc5RoC7QIQAvD_BwE#tab=tab_1)
2. *What Are the Different Types of Diabetes?* (https://www.healthline.com/health/diabetes/types-of-diabetes#causes)
3. *International Diabetes Federation* (https://idf.org/)
4. *Diabetes Complications* (https://medlineplus.gov/diabetescomplications.html)
5. Anwar, F., Ejaz, M. Y., & Mosavi, A. (2020). A comparative analysis on diagnosis of diabetes mellitus using different approaches: A survey. *Informatics in Medicine Unlocked*, *21*, 100482.

6. Sun, Y. L., & Zhang, D. L. (2019). Machine learning techniques for screening and diagnosis of diabetes: A survey. *Tehnički vjesnik, 26*(3), 872–880.
7. Alqahtani, N., Khan, W. A. G., Alhumaidi, M. H., & Ahmed, Y. A. A. R. (2013). Use of glycated hemoglobin in the diagnosis of diabetes mellitus and pre-diabetes and role of fasting plasma glucose, oral glucose tolerance test. *International Journal of Preventive Medicine, 4*(9), 1025.
8. Ang, S. H., Thevarajah, M., Alias, Y., & Khor, S. M. (2015). Current aspects in hemoglobin A1c detection: A review. *Clinica Chimica Acta, 439*, 202–211.
9. Pal, S., Mishra, N., Bhushan, M., Kholiya, P. S., Rana, M., & Negi, A. (2022, March). Deep learning techniques for prediction and diagnosis of diabetes mellitus. In *2022 International Mobile and Embedded Technology Conference (MECON)*, Noida, India (pp. 588–593). IEEE.
10. Lee, J. E., Jeon, H. J., Lee, O. J., & Lim, H. G. (2024). Diagnosis of diabetes mellitus using high frequency ultrasound and convolutional neural network. *Ultrasonics, 136*, 107167.
11. Dénes-Fazakas, L., Simon, B., Hartvég, Á., Kovács, L., Dulf, É. H., Szilágyi, L., & Eigner, G. (2024). Physical activity detection for diabetes mellitus patients using recurrent neural networks. *Sensors, 24*(8), 2412.
12. Kaul, D., Raju, H., & Tripathy, B. K. (2022). Deep learning in healthcare. In Acharjya, D. P., Mitra, A., & Zaman, N. (eds.), *Deep Learning in Data Analytics: Recent Techniques, Practices and Applications*, pp. 97–115. Cham: Springer.
13. Chauhan, C., & Gullapalli, R. R. (2021). Ethics of AI in pathology: Current paradigms and emerging issues. *The American Journal of Pathology, 191*(10), 1673–1683.
14. Islam, M. M., Yang, H. C., Poly, T. N., Jian, W. S., & Li, Y. C. J. (2020). Deep learning algorithms for detection of diabetic retinopathy in retinal fundus photographs: A systematic review and meta-analysis. *Computer Methods and Programs in Biomedicine, 191*, 105320.
15. Li, F., Wang, Y., Xu, T., Dong, L., Yan, L., Jiang, M., Zhang, X., Jiang, H., Wu, Z., & Zou, H. (2022). Deep learning-based automated detection for diabetic retinopathy and diabetic macular oedema in retinal fundus photographs. *Eye, 36*(7), 1433–1441.
16. Yang, H., Li, J., Liu, S., Yang, X., & Liu, J. (2022). Predicting risk of hypoglycemia in patients with type 2 diabetes by electronic health record–based machine learning: Development and validation. *JMIR Medical Informatics, 10*(6), e36958.
17. Cheng, Y., Wang, F., Zhang, P., & Hu, J. (2016, June). Risk prediction with electronic health records: A deep learning approach. In *Proceedings of the 2016 SIAM International Conference on Data Mining* (pp. 432–440), Miami, FL. Society for Industrial and Applied Mathematics.
18. Zhu, T., Li, K., Herrero, P., & Georgiou, P. (2020). Deep learning for diabetes: A systematic review. *IEEE Journal of Biomedical and Health Informatics, 25*(7), 2744–2757.

19. Zhu, T., Uduku, C., Li, K., Herrero, P., Oliver, N., & Georgiou, P. (2022). Enhancing self-management in type 1 diabetes with wearables and deep learning. *NPJ Digital Medicine*, *5*(1), 78.
20. Wong, K. K., Fortino, G., & Abbott, D. (2020). Deep learning-based cardiovascular image diagnosis: A promising challenge. *Future Generation Computer Systems*, *110*, 802–811.
21. Guan, Z., Li, H., Liu, R., Cai, C., Liu, Y., Li, J., Wang, X., Huang, S., Wu, L., Liu, D., Yu, S., Wang, Z., Shu, J., Hou, X., Yang, X., Jia, W., & Sheng, B. (2023). Artificial intelligence in diabetes management: Advancements, opportunities, and challenges. *Cell Reports Medicine*, *4*(10), 101213.
22. Nadeem, M. W., Goh, H. G., Hussain, M., Liew, S. Y., Andonovic, I., & Khan, M. A. (2022). Deep learning for diabetic retinopathy analysis: A review, research challenges, and future directions. *Sensors*, *22*(18), 6780.

CHAPTER 2

Preprocessing and Detection of Diabetes Mellitus from Physiological Data Using Deep Learning

Muskan Sah, Prajwal Kulkarni,
Pranav Sehgal, and Akila Victor

2.1 INTRODUCTION

Given that its incidence has been gradually increasing over the past few decades, diabetes mellitus (DM) poses a serious threat to global health. The International Diabetes Federation (IDF) estimates that 643 million adults worldwide, aged 20–79, will have diabetes (DM) by 2030, representing an increase from the estimated 537 million in 2021 [1]. This underscores the critical need for efficient methods in the diagnosis, prognosis, and treatment of DM. Diabetes mellitus (DM) is a collection of metabolic diseases marked by prominent blood sugar levels brought on by deficiencies in either insulin production, insulin function, or both [2]. If diabetes is not controlled, it can result in serious side effects such as neuropathy,

DOI: 10.1201/9781032708430-2

nephropathy, cardiovascular disease, and retinopathy, which can be very costly for both people and healthcare systems [3].

Early detection and management of diabetes are essential for preventing complications and improving patient results. While blood tests and clinical evaluations are commonly used in traditional techniques of diagnosing DM, they may not always be able to identify subtle early-stage signs or risk factors. As a result, increasing attention is being paid to utilizing cutting-edge computational methods, including deep learning, to improve the prognostic potential of DM detection models.

This research uses the Behavioural Risk Factor Surveillance System (BRFSS) dataset to investigate the application of deep learning techniques for preprocessing DM data and enhancing DM prognosis [4]. One of the world's largest health-related phone survey systems, the BRFSS, provides a plethora of information about different health behaviours, chronic illnesses, and the use of preventative services by a wide range of American demographics. This research builds upon existing research by investigating the efficacy of a deep learning approach using the Keras library with a TensorFlow backend for detecting DM. The dataset comprising 25 features selected from 303 unique variables, with the diabetes column serving as the target variable, is cleaned, balanced, and pre-processed to ensure data quality and balance.

By adopting feature selection and missing value imputation methods that can leverage the combination of large-scale health data and state-of-the-art deep learning algorithms, we want to make a positive impact on the rapidly developing field of computational healthcare through this research.

2.2 PROPOSED METHOD

The proposed methodology involves several steps as shown in Figure 2.1, starting with data preprocessing to clean and balance the dataset. A sequential deep learning model with validation for binary classification is constructed using the Keras library with a TensorFlow backend. The model architecture consists of fully connected layers with appropriate activation functions. The dataset is split into training and testing sets, with an additional validation split applied during model training to monitor for overfitting. It is trained on the training set and evaluated using various performance metrics.

The steps involved in the implementation are as follows:

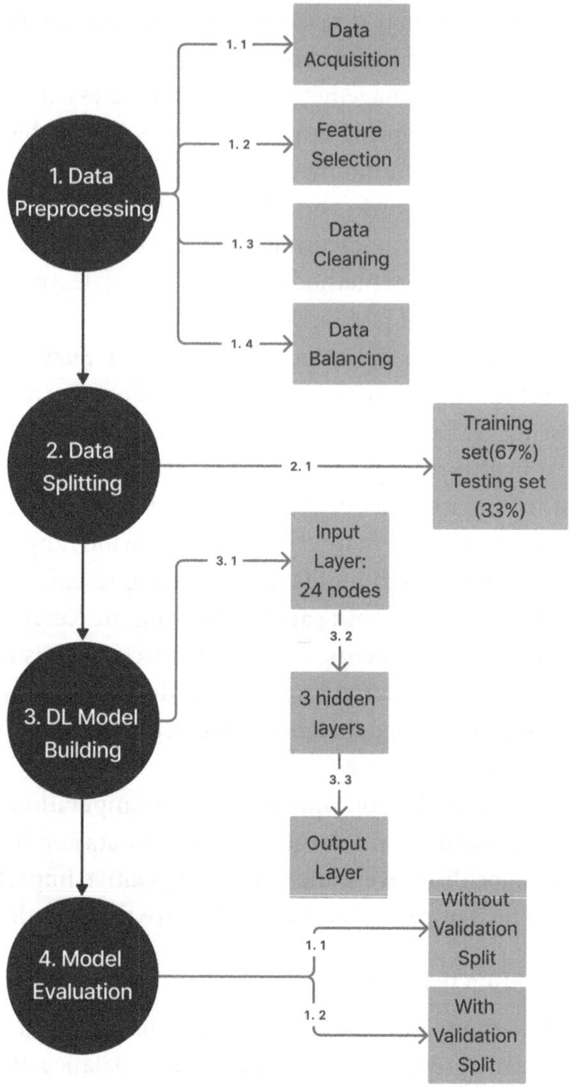

FIGURE 2.1 Flowchart of the proposed model.

2.2.1 Data Acquisition

The Centres for Disease Control and Prevention (CDC) conducts a large-scale annual telephone survey called the Behavioural Risk Factor Surveillance System (BRFSS), which is where the dataset used in this study originated. The survey collects responses from over 400,000 Americans on health-related risk behaviours, chronic health conditions, and the use of preventative services which has been conducted every year since 1984.

	_STATE	FMONTH	IDATE	IMONTH	IDAY	IYEAR	DISPCODE	SEQNO	_PSU	CTELENM1
0	1	1	1192021	1	19	2021	1100	2021000001	2021000001	1.0
1	1	1	1212021	1	21	2021	1100	2021000002	2021000002	1.0
2	1	1	1212021	1	21	2021	1100	2021000003	2021000003	1.0
3	1	1	1172021	1	17	2021	1100	2021000004	2021000004	1.0
4	1	1	1152021	1	15	2021	1100	2021000005	2021000005	1.0
...	...	...	...	...	...	...	...	...	...	...
438688	78	12	1062022	1	6	2022	1100	2021001381	2021001381	NaN
438689	78	12	1122022	1	12	2022	1100	2021001382	2021001382	NaN
438690	78	12	12212021	12	21	2021	1100	2021001383	2021001383	NaN
438691	78	12	1112022	1	11	2022	1100	2021001384	2021001384	NaN
438692	78	12	12222021	12	22	2021	1100	2021001385	2021001385	NaN

FIGURE 2.2 BRFSS dataset used in model implementation [5].

This original dataset has '303' features and responses from '438,693' people. These features are either questions that are distinctly asked to participants or variables that are processed based on the responses from certain participants (Figure 2.2).

2.2.2 Feature Selection

A total of 25 features were meticulously chosen from the original 303 features based on their potential relevance to diabetes prediction [5]. The following are significant risk factors for diabetes and other chronic illnesses like heart disease, according to research in the field (not in a strict order of importance) (Table 2.1).

2.2.3 Data Preprocessing

The data preprocessing phase involves several steps to ensure data quality and consistency. The steps applied were compiled from a BRFSS 2015 Codebook implementation [6]. Missing values are handled by using a function called *dropna()* which removes rows or columns with missing values (NaN) from a data frame. Duplicates are removed to prevent redundancy, and categorical variables are encoded or transformed as needed. Additionally, continuous variables may be standardized or normalized to ensure uniform scaling across features.

2.2.4 Data Balancing

The number of negative samples (people without diabetes) in the original dataset was higher than the number of positive samples (people with diabetes), indicating a class imbalance. To create a balanced dataset, a 50–50 split between positive (diabetes) and negative (no diabetes) samples is achieved

TABLE 2.1 Features Selected from BRFSS Dataset

	Feature Name	Description
1.	Diabetes_binary	Diabetes awareness
2.	BMI	Body mass index
3.	Overweight	Body mass index (BMI) greater than 25.00 (overweight or obese)
4.	High BP	High blood pressure awareness
5.	High cholesterol	Cholesterol awareness
6.	Cholesterol check	Cholesterol checks within past 5 years
7.	Kidney problem	Unable to control urine flow or kidney disease
8.	Stroke	Smoked at least 100 cigarettes
9.	Alcoholic	Heavy alcohol consumption
10.	Chronic health problem	Has had a stroke before
11.	Heart problem	Ever has coronary heart disease (CHD) or myocardial infarction (MI)
12.	Physical activity	Does physical activity or exercise
13.	Eat fruits	Consume fruit at least once per day
14.	Eat vegetables	Consume vegetable at least once per day
15.	General health	General health status
16.	Physical health	The number of days physical health is good
17.	Mental health	The number of days mental health is not good
18.	Difficulty walking	Difficulty walking or climbing stairs
19.	Healthcare access	Have any healthcare access or insurance
20.	Not wealthy to visit a doctor	Couldn't afford to see a doctor
21.	Regular checkup	Last routine checkup
22.	Sex	Male or female
23.	Age	Reported age in 5-year age categories. It starts with 1 for age 18–24 all the way to 13 for 80 years and older
24.	Education	Highest education level achieved
25.	Income	Annual income level

through a resampling technique called under-sampling. Under-sampling involves reducing the size of the majority class (in this case, samples with no diabetes, labelled as 0) to match the size of the minority class (samples with pre-diabetes and diabetes, labelled as 1 and 2). By randomly selecting a subset of the majority class instances, the resulting dataset achieves a more balanced distribution between the classes. In this specific case, both the no-diabetes and diabetes classes are down sampled to 33,289 instances each, resulting in a balanced dataset.

2.2.5 Model Architecture

The chosen deep learning model follows a sequential architecture using Keras library comprising of an input layer, two hidden layers, and an output layer. The input layer has 64 neurons with an input dimension of 24 corresponding to the selected features, while the hidden layers use rectified linear unit (ReLU) activation functions to introduce non-linearity into the model. Two additional hidden layers are added to the model. The first hidden layer has 32 neurons, and the second hidden layer has 20 neurons with ReLU activation to help the model learn complex patterns and relationships within the data. A final sigmoid activation function with one neuron in the output layer produces a binary classification output, indicating the likelihood of the presence of diabetes (Figure 2.3).

2.2.6 Model Compilation and Training

In Keras is the process of preparing the model for training. The compilation includes the definition of several important parameters.

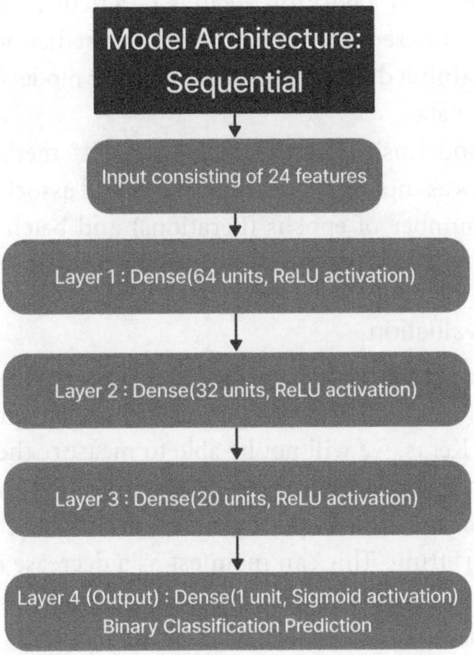

FIGURE 2.3 DL model architecture.

- **Optimizer:** The optimizer is the algorithm that is used to update the weights of the neural network during training to minimize the loss function. 'Adam' is a popular optimizer in DL.
- **Loss function:** The loss function quantifies the difference between the predicted values and the actual values showing how well the model performs on the training data and guides the optimization process. For binary classification problems like this one, 'binary_crossentropy' is commonly used.
- **Metrics:** Metrics are used for evaluating the performance of the model during training and testing. 'Accuracy' is used, which measures the proportion of correctly classified samples.

The model is trained twice. The former is split into testing (67%) and training (33%), and the latter is split into training and validation. Model training involves adapting the model weights to the training dataset. This entails repeating multiple epochs of training the model with the training data. This is done by using a cost function and minimizing this function using an optimization algorithm such as Gradient Descent. The goal is to minimize the difference between the model predictions and the actual values for the training dataset. After training, the model can make predictions on unseen data.

To train a model using Keras, we use the "fit" method of the model object. This takes as input the training dataset and associated target labels, as well as the number of epochs (iterations) and batch size (number of examples used to update the model weights each time).

2.2.7 Model Evaluation

It entails calculating the model's accuracy and loss, among other measures, to gauge its performance. If we do not use validation split when training the model with Keras, we will not be able to measure the performance of the model on an independent dataset during training. This can lead to an overestimation of the model's performance on the training data, which is often called overfitting. This can manifest as a decrease in the loss of the training data over epochs. This means that the model starts memorizing the training data instead of generalizing the relationships to new data. It is therefore important to use validation split to evaluate the performance of the model on independent data during training.

Model performance was evaluated on both training and testing sets using the evaluate method:

- **Training accuracy:** Measured to assess model fit to the training data.
- **Testing accuracy:** Measured to assess model generalizability to unseen data.

2.3 RESULT AND ANALYSIS

Various metrics, including accuracy, precision, recall, and F1-score, were used to assess the performance of the model. Without the validation split, the training accuracy was determined to be 75.34%, while the testing accuracy reached 74.21%. The precision, recall, and F1-score were all calculated to be 0.7420, with a total support of 21,971 instances. With the validation split, the training accuracy slightly decreased to 74.34%, while the validation accuracy increased to 74.38%. Similarly, the precision, recall, and F1-score were all consistent at 0.7437, with a support of 21,971 instances. The confusion matrix with the validation split parameter revealed that the model had a higher precision for predicting negative instances (0.81) compared to positive instances (0.70), while recall was higher for positive instances (0.85) compared to negative instances (0.64). These metrics collectively demonstrate the model's ability to accurately classify instances of DM, highlighting its effectiveness in disease detection and risk assessment.

The validation split is crucial for evaluating the model's performance on an independent dataset. Without it, the model may overfit, resulting in a decrease in the loss of training data over epochs. Therefore, it's essential to use validation split to measure the model's performance on the independent dataset, preventing overfitting and ensuring accurate training data evaluation (Figures 2.4 and 2.5).

2.4 DISCUSSION

The results of the deep learning model trained to predict DM based on physiological data exhibit promising performance metrics. Both scenarios, with and without the validation split, achieved comparable accuracies, indicating the robustness of the model. The precision, recall, and F1-score metrics further validate the effectiveness of the model in correctly identifying instances of DM. However, the model exhibited a small decrease in accuracy from training to testing, suggesting some degree of overfitting. This implies that the model might have memorized

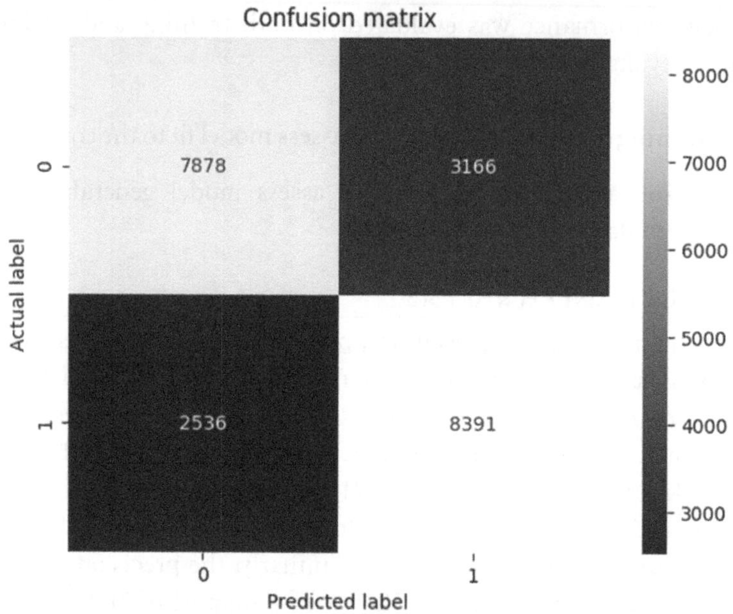

FIGURE 2.4 Confusion matrix for the model without validation [5].

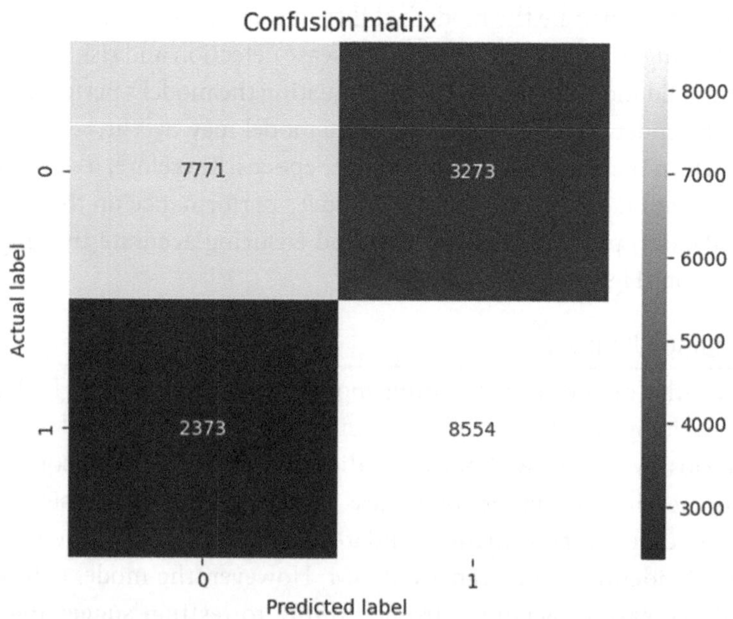

FIGURE 2.5 Confusion matrix for the model with validation [5].

the training data patterns too well and may not generalize effectively to unseen data. The confusion matrix illustrates the distribution of correct and incorrect predictions, highlighting areas where the model excels and potential areas for improvement. Furthermore, changing hyperparameters, adding or removing layers, and other model adjustments may be required for optimization to achieve results with more promising accuracy.

According to the results, the model may be able to accurately identify DM from BRFSS data. Nonetheless, there exist prospects for additional refinement and enhancement of the model. To improve the model's functionality and generalizability, more feature engineering approaches, different model architectures, and bigger datasets may be investigated in subsequent studies. Furthermore, adding domain-specific information and clinical experience could enhance the model's interpretability and clinical value. This approach ensures robustness and generalizability in the detection of DM using deep learning techniques (Figures 2.6–2.8).

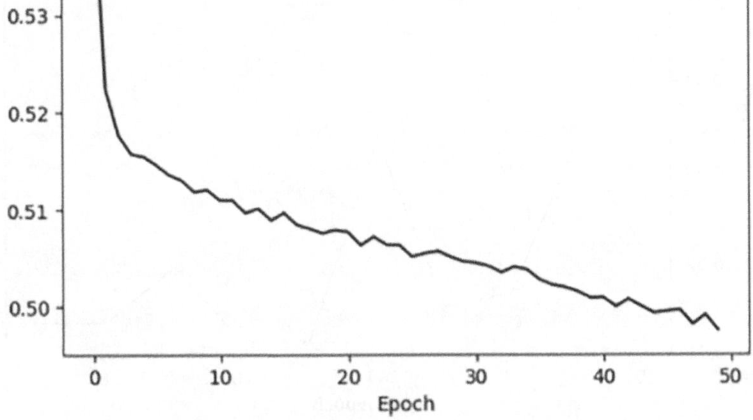

FIGURE 2.6 Visual analysis of loss of training without validation [5].

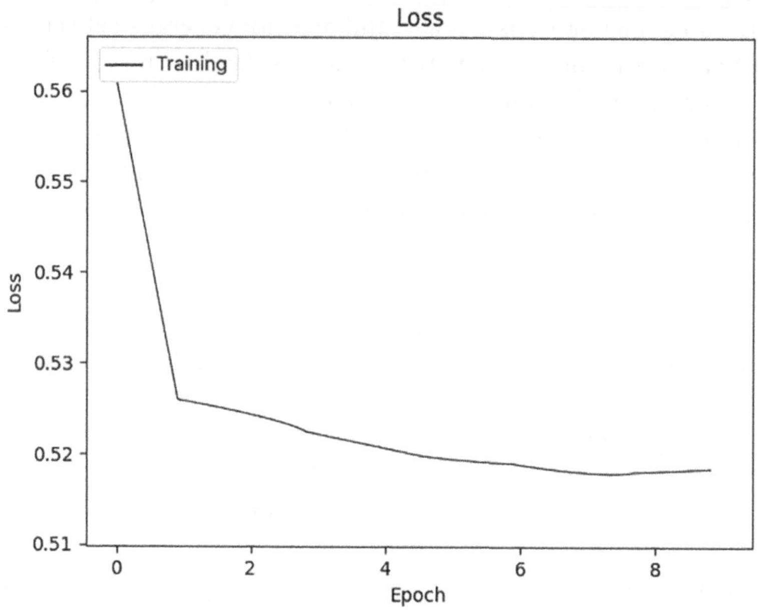

FIGURE 2.7 Visual analysis of training loss [5].

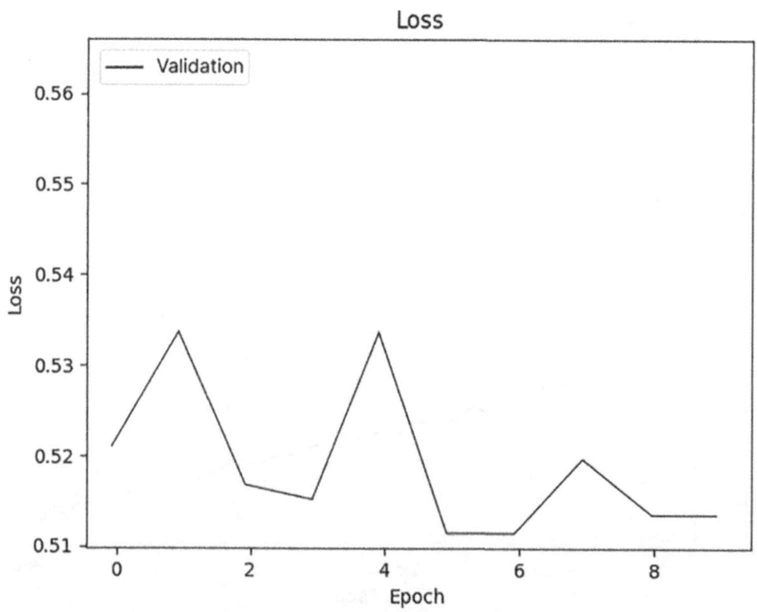

FIGURE 2.8 Visual analysis of validation loss [5].

2.5 CONCLUSION

With approximately 74% training accuracy and 74% testing accuracy, the model with validation split parameter demonstrated a reasonable level of accuracy, indicating its potential utility as a screening tool for preliminary information. The sequential deep learning model with validation for binary classification effectively constructed a deep learning model to identify DM using physiological data. Strong performance measures were demonstrated by the model, including high values for recall, accuracy, precision, and F1-score. The model's ability to rectify class imbalances and produce reliable predictions was greatly enhanced using a balanced dataset and careful preprocessing methods. These results demonstrate the great promise of deep learning approaches in the healthcare domain, especially in applications related to risk assessment and disease detection. To improve the model's functionality and make it easier for it to be integrated into clinical practice, future research may investigate optimizing techniques and adding more features.

REFERENCES

1. International Diabetes Federation. (2021). *IDF Diabetes Atlas*, 10th edn. Brussels, Belgium: International Diabetes Federation.
2. American Diabetes Association. (2022). Diagnosis and classification of diabetes mellitus. *Diabetes Care*, 45(Suppl 1), S17–S38.
3. American Diabetes Association. (2022). Standards of medical care in diabetes: 2022. *Diabetes Care*, 45(Suppl 1), S1–S236.
4. https://www.cdc.gov/brfss/annual_data/2021/pdf/codebook21_llcp-v2-508.pdf
5. https://www.kaggle.com/code/muskansah/diabetes-mellitus-detection
6. https://www.kaggle.com/code/hakonem/diabetes-health-indicators-dataset-notebook

CHAPTER 3

Graph-Based Explainable Method for Blood Glucose Prediction through Federated Learning

Chengzhe Piao and Kezhi Li

3.1 INTRODUCTION

Type 1 diabetes (T1D) occurs if the pancreas fails to generate sufficient insulin, resulting in elevated blood glucose (BG) levels (hyperglycemia, defined as BG > 180 mg/dL). To manage their condition and maintain BG levels within the normal range (70–180 mg/dL), individuals with T1D rely on insulin delivery methods such as pumps and bolus injections. Without proper management, there is a significant risk of experiencing hypoglycemia (BG < 70 mg/dL) or hyperglycemia, which can lead to severe complications, including kidney issues, vision problems, and even blindness.

To prevent these serious health issues, it is crucial for people with T1D to actively monitor and manage their BG levels. Continuous glucose monitoring (CGM) systems have become a vital tool in this regard, offering

DOI: 10.1201/9781032708430-3

the ability to track BG level fluctuations every 5 minutes. Leveraging the detailed data provided by CGM, researchers have developed a variety of predictive models for blood glucose level prediction (BGLP). These models enable individuals with T1D to proactively adjust their insulin dosages and dietary intake, significantly enhancing their ability to maintain stable BG levels and mitigate the risk of complications.

In BGLP, historical data is often structured as multivariate time series (MTS) to incorporate various attributes relevant to the prediction task. Since 2020, neural network (NN)-based methodologies have demonstrated superior performance over traditional non-NN approaches, particularly highlighted during the BGLP 2020 Challenge. Among these advancements, Neural Basis Expansion Analysis for Time Series (N-BEATS) (Oreshkin et al., 2020) has been applied by Rubin-Falcone et al. (2020). N-BEATS was a leading approach for time-series analysis at the time, and their N-BEATS-based solution notably secured first place in the challenge. Despite these achievements, a significant limitation remains: These models offer limited insight into the significance of different attributes, such as glucose levels, bolus, and finger sticks, in the prediction process. This lack of attribute importance explanation is a gap not only in N-BEATS but also in other contemporary BGLP methodologies that rely on recurrent neural network (RNN) architectures like long short-term memory (LSTM; Hochreiter and Schmidhuber, 1997; Yang et al., 2020). These approaches can capture temporal dependencies but fall short in transparently modeling the interconnections among attributes. The importance of this limitation cannot be understated, as many clinicians and healthcare professionals prioritize understanding the decision-making process of predictive models. The ability to interpret how different attributes influence predictions is crucial for clinical acceptance and trust in these technologies.

Therefore, a critical limitation of these approaches is their inability to quantify and explain the significance of each attribute within the BGLP context. To address this gap, advancements in graph neural networks (GNNs) present a promising avenue, particularly those employing dynamic graphs with adaptable edge weights. Graph attention networks (GATs; Velickovic et al., 2018) enable the edge weights, predicated on the adjacent node tensors, to be adjustable during the learning process. Subsequently, Brody et al. (2022) developed an enhanced version, GATv2, which incorporates a more dynamic attention mechanism, offering increased model expressiveness. These innovations in GAT and GATv2 facilitate the explanation of attribute significance in BGLP through the modulation of edge weights.

In this research, our proposed model is grounded in the original GAT framework (see Section 3.2). Although GATv2 presents certain advantages, our preliminary experiments did not yield significant improvements to justify its adoption over GAT.

In terms of NN training and deployment for BGLP, recent innovations have emphasized the development of personalized models through multitask learning, as highlighted by Daniels et al. (2020, 2022). This approach diverges from traditional methods that required the aggregation of individual data to train a generalized population model, which was subsequently fine-tuned with personal data. Multitask learning facilitates the simultaneous capture of population-wide patterns and individual-specific characteristics, thus enabling the generation of personalized models without extensive fine-tuning. As for the deployment, Zhu et al. (2021) demonstrated the feasibility of deploying an LSTM-based model on a microcontroller unit for real-time BG prediction in individuals with T1D. Further extending the technological frontier, Zhu et al. (2022) employed a wearable device, incorporating a low-cost chip capable of running a temporal attention-based GRU model. This approach leverages desktop and cloud platforms for model fine-tuning and data storage, showcasing a practical application of NNs in wearable health monitoring devices. Additionally, D'Antoni et al. (2022) utilized TensorFlow and Keras API tools to implement an LSTM-based model directly on edge devices. This method slightly compromises prediction accuracy for the benefit of real-time processing capabilities, which is acceptable within the context of their application. However, these methodologies often overlook the crucial aspect of privacy protection during the aggregation of data for learning population patterns. This oversight can limit participant recruitment and consequently restrict the volume of available training data. Addressing this challenge remains critical for advancing NN-based BGLP methods, ensuring both effective prediction capabilities and adherence to privacy standards.

Hence, a notable deficiency in recent BGLP methodologies is the lack of consideration for privacy protection during the training of population models. Federated learning (FL), characterized by its decentralized learning approach, emerges as a solution to this challenge. FL aims to harness the computational capabilities of distributed devices for artificial intelligence (AI) training tasks directly on the data produced by these devices, thereby maintaining data confidentiality at its source. The model updates are aggregated on a central server through a process known as federated averaging (FedAvg), as proposed by McMahan et al. (2017), ensuring that only model parameters are shared rather than sharing data. Given FL's potential for safeguarding participant privacy while facilitating the learning of population-wide patterns, this work explores the application of FL within the BGLP domain. Our goal is to assess the feasibility and effectiveness of integrating FL in the training of deep models, aiming to strike a balance between predictive accuracy and privacy preservation (see Section 3.2).

In our experiments (see Section 3.3), we designed three experiments in OhioT1DM (Marling and Bunescu, 2018, 2020). The initial trial established that the contributory factors include "glucose level", "bolus", "meal", "exercise", "finger stick", and "sleep". In subsequent experiments, variations in hyperparameters of our methods were examined, demonstrating its stability across diverse setups. The third experiment delved into the impact of FL, confirming a reduction in prediction accuracy when integrated with our approach. Finally, our discussion and conclusion (see Sections 3.4 and 3.5) are given at the end of this chapter.

3.2 METHOD

3.2.1 Problem Definition

BGLP via regular multivariate time series (RMTS-BGLP): If we have N factors from previous T time steps, the data can be represented as $\mathbf{X}=[\mathbf{x}_1 \cdots \mathbf{x}_T]$, where $\mathbf{X} \in \mathrm{R}^{N \times T}$. Note that $\mathbf{x}_t$ has N factors. The task is to forecast the BG level y_{T+W} in the future. We set the prediction horizon W to 6 or 12, representing the prediction in 30/60 minutes. Regarding $T \triangleq \{1, \ldots, T\}$ and aligning with the sampling frequency of CGM, each adjacent pair of the time series is 5 minutes.

3.2.2 Setting and Dataset

This study examines two datasets: OhioT1DM'18 and OhioT1DM'20 (Marling and Bunescu, 2018, 2020). There are 12 participants, with six from each dataset, all diagnosed with T1D. Details of these datasets are provided in Table 3.1 with key observations as follows: (i) The participant group includes five females and seven males, offering a relatively balanced gender distribution; (ii) the age range for the majority of participants is middle-aged, with eight participants falling within the 40–60 age bracket; and (iii) for the majority of participants, the ratio of training to testing samples is approximately 5:1.

TABLE 3.1 Basic Information of the OhioT1DM (Marling and Bunescu, 2020)

Patient ID (PID)	BGLP Challenge	Gender	Age	#Training Examples	#Test Examples	No. of Days
559	2018	Female	40–60	10,796	2514	52
563	2018	Male	40–60	12,124	2570	56
570	2018	Male	40–60	10,982	2745	51
575	2018	Female	40–60	11,866	2590	56
588	2018	Female	40–60	12,640	2791	56
591	2018	Female	40–60	10,847	2760	53
540	2020	Male	20–40	11,947	2884	57
544	2020	Male	40–60	10,623	2704	52
552	2020	Male	20–40	9080	2352	48
567	2020	Female	20–40	10,858	2377	57
584	2020	Male	40–60	12,150	2653	57
596	2020	Male	60–80	10,877	2731	57

3.2.3 Ethics

The OhioT1DM dataset is available to researchers solely for research purposes (Marling and Bunescu, 2020). Consequently, no further ethics approvals are necessary to access the OhioT1DM'18 and OhioT1DM'20 datasets.

3.2.4 Proposed Method

In this study, we introduce a novel neural network model known as "GAM", which is enhanced by FL. Figure 3.1 illustrates that GAM incorporates a GRU, as detailed in Cho et al. (2014), and a GAT, as described in Velickovic et al. (2018). GAT is utilized to effectively capture the patterns among different factors in a RMTS, created from an irregular multivariate time series (IMTS) via zero padding. Each non-zero attribute value is treated as a dynamic graph node, transformed into a tensor through fully connected layers, thus forming graph embeddings. Graph features are extracted at each time slot through graph convolutions applied to these embeddings, using an adjacency matrix. These features are then temporally integrated by the GRU to produce a concise vector of temporal graph features, which is subsequently employed in predicting BG levels.

Specifically, the formation of $\hat{y}_{T+W} = \text{GAM}(X; \theta)$ can be represented as follows:

$$e_t^n = f_n\left(x_t^n\right) \tag{3.1}$$

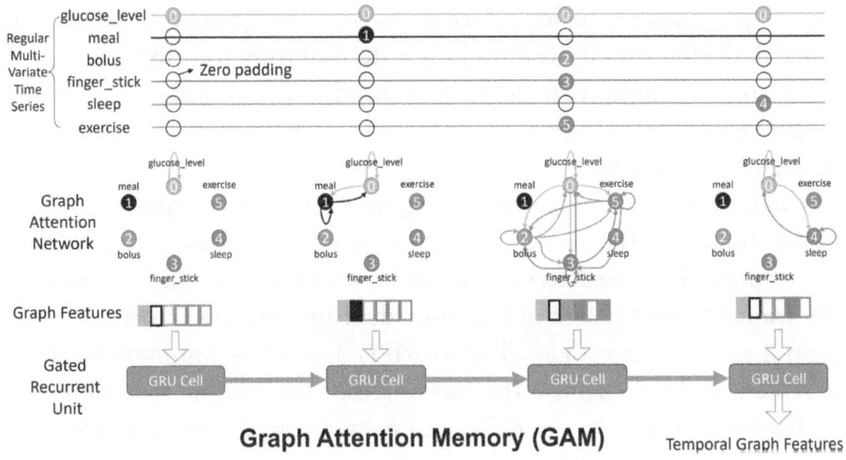

FIGURE 3.1 Graph attention memory (GAM).

$$\mathbf{E}_t = \left[\mathbf{e}^1{}_t \cdots \mathbf{e}^n{}_t \cdots \mathbf{e}^N{}_t\right] \tag{3.2}$$

$$\mathbf{G}_t = \text{GATs}(\mathbf{E}_t, \mathbf{A}), \tag{3.3}$$

$$\mathbf{h}_t = \text{GRU}\left(\phi(\mathbf{G}_t), \mathbf{h}_{t-1}\right), \tag{3.4}$$

$$\hat{y}_{T+W} = f_{\text{out}}(\mathbf{h}_T). \tag{3.5}$$

Let T be defined as the set $\{1,\ldots, t,\ldots, T\}$ and N be defined as the set $\{1,\ldots, n,\ldots, N\}$. Let x_t^n denote the element located in the t-th column and n-th row of matrix $\mathbf{X}$, where $\mathbf{X}$ belongs to the set of matrices with dimensions $R^{N\times T}$. The functions $f(\cdot)$ represent neural network layers. $\mathbf{A}$ is a graph adjacency matrix ($R^{N\times N}$). The hidden state of a GRU is $\mathbf{h}_t$, and $\mathbf{h}_0$ is initialized with zeros. The estimated value of y_{T+W} is $\hat{y}_{T+W}$.

In Equation (3.1), each non-zero item x_t^n of the matrix $\mathbf{X}$, which is part of the RMTS, undergoes individual embedding via $f_n(\cdot)$, with each factor $n \in N$ having its corresponding $f_n(\cdot)$. This tensor produces $\mathbf{e}_t^n \in R^E$, representing x_t^n, and is aggregated into $\mathbf{E}_t$, as shown in Equation (3.2). Subsequently, as illustrated in Equation (3.3), GATs are employed to perform graph convolutions on $\mathbf{E}_t$ using the adjacency matrix $\mathbf{A}$. Following the graph convolutions, $\mathbf{G}_t \in R^{E\times N}$ is produced, encapsulating the interrelations among all attributes. Afterward, $\mathbf{G}_t$ is transformed into a tensor using the function $\phi(\cdot)$, resulting in $\phi(\mathbf{G}_t)$. Subsequently, $\{\phi(\mathbf{G}_1), \ldots, \phi(\mathbf{G}_T)\}$ are aggregated by a GRU, producing a condensed tensor $\mathbf{h}_T$ containing temporal patterns. In the end, the forecasted BG level $\hat{y}_{T+W}$ is derived from $\mathbf{h}_T$ using $f_{\text{out}}(\cdot)$.

The mean square error (MSE) is utilized as the training loss, and the Adam optimizer (Kingma and Ba, 2015) is employed to optimize the learnable parameters.

Furthermore, FL is illustrated in Figure 3.2. Only the respective personalized models on the clients have access to personal data. After undergoing training for a predetermined number of epochs on the local devices, the trainable parameters of the individual method are transmitted to a central server, where a global approach is derived by summarizing these parameters. Subsequently, the personalized models on the clients are updated with the global model's latest parameters, effectively synchronizing the personal and global learnings. This cycle continues until the models reach an optimal state. Through this methodology, participant privacy is robustly safeguarded, because personal data remains exclusively within their domain. At the same time, the server is able to aggregate global trends without compromising individual privacy. It's worth noting that in our study, FL is emulated through a multi-process approach, treating each client and the server as separate processes.

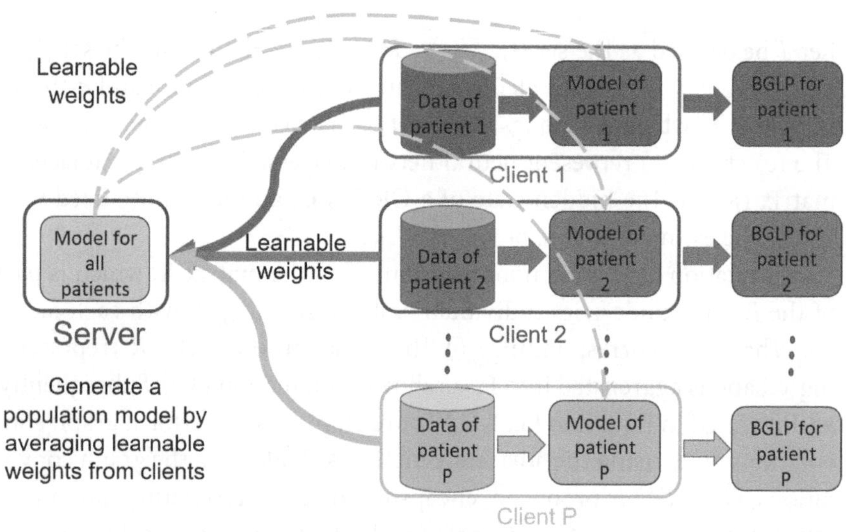

FIGURE 3.2 Federated learning (FL).

3.2.5 Metrics

We consider three metrics in this work. RMSE is employed to thoroughly assess model performance, guiding the selection of optimal trainable parameters and attribute combinations. Besides, mean absolute relative difference (MARD) is also considered, making it ideal for evaluating predictions of lesser magnitudes. Moreover, mean absolute error (MAE) is used for an easy comparison with other models, especially since MAE is a standard metric used by all entries in the Ohio 2020 challenge.

3.3 EXPERIMENTS AND RESULTS

3.3.1 Impact of Attributes

To effectively leverage various attributes in the OhioT1DM dataset and experimental resources, we experimented with a select number of attribute combinations. Interestingly, we discovered that employing six specific attributes, including "glucose level", "bolus", "meal", "sleep", "exercise", and "finger stick", consistently yields noticeable improvements in most scenarios.

Therefore, our initial experiment aimed to determine the positive influence of each of these six attributes on RMTS-BGLP. In this experiment, we systematically excluded one attribute at a time (apart from "glucose level") from the set of six attributes. We then compared the performance of GAM utilizing all six attributes against the performance of GRU using only BG. Next, if GAM with these six attributes outperforms the others, it indicates that each attribute contributes positively to RMTS-BGLP. Conversely, if GRU with only BG shows the least favorable outcomes, it suggests that the combination of these six attributes is effective. The results of this experiment are detailed in Tables 3.2 and 3.3.

Additionally, we aim to explain our decision to exclude certain attributes. To do this, we tested the inclusion of one additional attribute at a time to the existing set of six attributes. We then compared the results to the performance of GAM with just the original six attributes. If the addition of a new attribute does not lead to improvements over the GAM containing the six factors in most scenarios, we opted not to include other factors (see Tables 3.4 and 3.5).

In general, when the prediction window W is set to 6, the "GAM+6 attributes" model demonstrates superior performance in terms of both RMSE and MARD across various settings. The mean RMSE for "GAM+6 attributes" is 18.91 mg/dL (W=6), outperforming other models whose

TABLE 3.2 The Influence of "Glucose Level", "Bolus", "Meal", "Finger Stick", "Exercise", and "Sleep" (Measured by RMSE, $W=6$)

Method	**559**	**563**	**570**	**575**	**588**	**591**	**540**	**544**	**552**	**567**	**584**	**596**	**Average**
GAM+6 attributes w/o meal	19.51	17.84	16.94	22.52	17.32	21.04	21.20	16.26	16.26	21.57	22.23	16.82	19.13
GAM+6 attributes w/o bolus	19.74	18.00	16.74	22.40	17.08	20.89	21.70	15.62	16.70	21.98	23.13	16.40	19.20
GAM+6 attributes w/o finger stick	19.95	17.82	17.00	22.46	16.94	21.46	20.99	16.22	16.41	21.98	22.63	16.10	19.16
GAM+6 attributes w/o sleep	19.01	17.91	16.12	22.50	17.17	20.83	21.06	16.12	16.47	21.52	22.28	16.27	18.94
GAM+6 attributes w/o exercise	19.11	17.98	16.99	22.42	17.30	21.00	20.89	16.27	16.19	21.54	22.00	16.02	18.97
GAM+6 attributes	19.42	17.60	16.66	22.28	17.06	20.72	21.01	15.95	16.50	21.58	22.22	16.01	18.92
GRU glucose level	19.44	18.29	18.60	23.62	18.49	21.83	22.77	18.00	17.37	22.92	23.59	17.20	20.18

TABLE 3.3 The Influence of "Glucose Level", "Bolus", "Meal", "Finger Stick", "Exercise", and "Sleep" (Measured by MARD, $W=6$)

Method	**559**	**563**	**570**	**575**	**588**	**591**	**540**	**544**	**552**	**567**	**584**	**596**	**Average**
GAM+6 attributes w/o meal	8.83	7.75	5.38	9.74	7.55	12.17	10.87	7.74	9.20	10.91	10.91	8.78	9.15
GAM+6 attributes w/o bolus	8.89	7.75	5.37	9.55	7.49	12.03	11.09	7.49	9.36	11.11	11.47	8.59	9.18
GAM+6 attributes w/o finger stick	8.97	7.73	5.38	9.50	7.37	12.25	10.88	7.72	9.21	11.14	11.20	8.55	9.16
GAM+6 attributes w/o sleep	8.71	7.73	5.21	9.60	7.45	11.85	10.78	7.62	9.24	10.65	10.93	8.48	9.02
GAM+6 attributes w/o exercise	8.62	7.71	5.36	9.48	7.50	11.94	10.75	7.75	9.06	10.85	10.79	8.49	9.03
GAM+6 attributes	8.79	7.65	5.26	9.44	7.42	11.97	10.78	7.61	9.15	10.87	11.03	8.43	9.03
GRU glucose level	8.99	7.98	5.74	10.27	8.03	12.35	11.58	8.69	9.66	11.24	11.26	8.98	9.56

TABLE 3.4 The Influence of Leveraging Other Attributes (Measured by RMSE, $W=6$)

Method	**559**	**563**	**570**	**575**	**588**	**591**	**540**	**544**	**552**	**567**	**584**	**596**	**Average**
GAM+6 attributes+basal	19.23	19.76	19.42	22.34	17.56	20.92	20.69	16.06	16.86	21.84	25.07	16.01	19.64
GAM+6 attributes+basis skin temperature	19.27	17.95	16.21	22.71	16.99	21.13	20.96	16.42	16.49	21.74	22.61	16.01	19.04
GAM+6 attributes+basis GSR	19.40	18.03	16.60	22.45	17.09	21.22	20.77	16.34	16.63	21.72	22.68	16.20	19.09
GAM+6 attributes+basis sleep	19.06	17.78	16.54	22.07	17.23	20.83	21.28	15.82	16.44	21.45	22.00	15.93	18.87
GAM+6 attributes+acceleration	19.31	17.93	16.42	22.24	17.21	20.71	21.08	15.84	16.69	21.82	22.15	15.89	18.94
GAM+6 attributes+work	19.39	17.78	16.80	22.10	16.94	20.83	21.10	16.06	16.54	21.50	21.75	15.91	18.89
GAM+6 attributes+basis steps	19.05	17.68	16.43	22.38	17.26	20.77	21.17	15.85	16.36	21.72	22.18	16.04	18.91
GAM+6 attributes+basis heart rate	19.67	17.44	16.71	22.57	16.97	20.62	21.13	16.10	16.52	21.65	22.04	16.05	18.96
GAM+6 attributes+basis air temperature	19.31	17.68	16.23	22.32	17.18	20.72	21.03	15.98	16.16	21.66	22.06	16.13	18.87
GAM+6 attributes+day of week	18.85	17.68	16.74	22.69	16.98	20.73	21.08	15.82	16.23	21.90	22.49	16.32	18.96
GAM+6 attributes+total seconds	19.24	17.87	16.32	22.33	16.90	21.17	20.90	16.07	16.53	21.88	22.50	16.09	18.98
GAM+6 attributes	19.42	17.60	16.66	22.28	17.06	20.72	21.01	15.95	16.50	21.58	22.22	16.01	18.92

TABLE 3.5 The Influence of Leveraging Other Attributes (Measured by MARD, $W=6$)

Method	559	563	570	575	588	591	540	544	552	567	584	596	Average
GAM+6 attributes+basal	8.75	8.51	7.15	9.48	7.71	12.05	10.70	7.64	9.40	10.99	13.94	8.36	9.56
GAM+6 attributes+basis skin temperature	8.67	7.68	5.29	9.81	7.38	12.05	10.84	7.76	9.38	10.74	11.10	8.47	9.10
GAM+6 attributes+basis GSR	8.88	7.76	5.27	9.85	7.51	12.00	10.76	7.65	9.46	10.86	11.16	8.52	9.14
GAM+6 attributes+basis sleep	8.47	7.71	5.28	9.36	7.46	11.80	10.84	7.59	9.11	10.55	10.96	8.40	8.96
GAM+6 attributes+acceleration	8.66	7.69	5.22	9.49	7.42	11.92	10.71	7.55	9.31	11.03	10.95	8.43	9.03
GAM+6 attributes+work	8.73	7.71	5.31	9.47	7.35	11.93	10.77	7.57	9.28	10.83	10.74	8.40	9.01
GAM+6 attributes+basis steps	8.64	7.63	5.27	9.49	7.44	11.85	10.77	7.50	9.16	10.87	10.92	8.46	9.00
GAM+6 attributes+basis heart rate	8.78	7.67	5.30	9.68	7.42	11.77	10.79	7.56	9.21	10.83	10.89	8.50	9.03
GAM+6 attributes+basis air temperature	8.61	7.68	5.42	9.56	7.43	11.87	10.73	7.57	9.12	10.91	10.83	8.54	9.02
GAM+6 attributes+day of week	8.59	7.69	5.32	9.54	7.37	11.96	10.83	7.53	9.14	10.99	11.17	8.53	9.06
GAM+6 attributes+total seconds	8.78	7.70	5.20	9.72	7.40	12.07	10.71	7.57	9.26	10.98	11.07	8.45	9.08
GAM+6 attributes	8.79	7.65	5.26	9.44	7.42	11.97	10.78	7.61	9.15	10.87	11.03	8.43	9.03

RMSE values exceed this benchmark (refer to Table 3.2). Conversely, "GRU glucose level" consistently exhibits the poorest performance across all scenarios. Specifically, the average RMSE for "GRU glucose level" is 20.17 mg/dL (W=6; see Table 3.2). Consequently, we infer that each of the six attributes positively contributes to the performance of RMTS-BGLP, and their collective integration yields significant improvements. Concerning additional attributes, as depicted in Tables 3.4 and 3.5, it becomes apparent that including new attributes generally does not lead to performance enhancements in most cases.

3.3.2 Impact of Hyperparameters

We choose two significant hyperparameters to assess their impact, namely H (the hidden size) of the GRU-based memory and the L (number of layers) of GAT. Then, H determines GAM's memory capacity, with a larger H enhancing its memory capabilities up to a certain threshold. In this experiment, we evaluate five different hidden sizes, specifically $H \in \{64, 128, 256, 512, 1024\}$.

Likewise, varying L in GAT influences GAM's capacity to represent data differently. Increasing L enhances GAM's processing capability. For instance, when L=1, the node representing "glucose level" can solely receive messages, and the message senders are its adjacent nodes. However, if we set L to 2, while the message exchange remains similar to L=1 in the first layer, the received messages are not the original graph embeddings in the subsequent layer. Instead, they are amalgamated before aggregation. Consequently, deeper layers enable the merging of graph messages in a more intricate manner, akin to extending the number of layers in convolutional neural networks (CNNs). In this experiment, we explore five different numbers of layers.

The influence of H is illustrated in Figure 3.3. It is evident that GAM achieves optimal performance with H=256 and W=6. Larger values of H, such as H>512, do not yield improved prediction. However, when increasing L, the mean RMSE improves from 19.02 mg/dL (L = 1) to 18.85 mg/dL (L = 4), as shown in Figure 3.4a. Therefore, it is inferred that, in this scenario, enhancing the capability of GAT by increasing L does not result in significant improvements compared to varying H.

3.3.3 Impact of FL

This experiment aims to evaluate the impact of incorporating FL on predictive capability during training. We compare the prediction capability

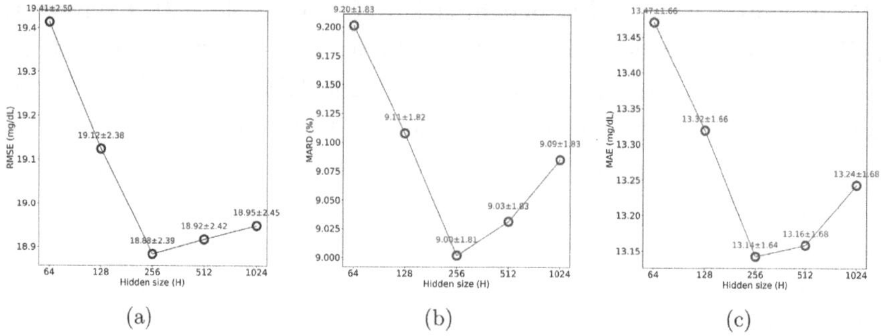

FIGURE 3.3 The influence of varying hidden size (H), prediction window (W=6).

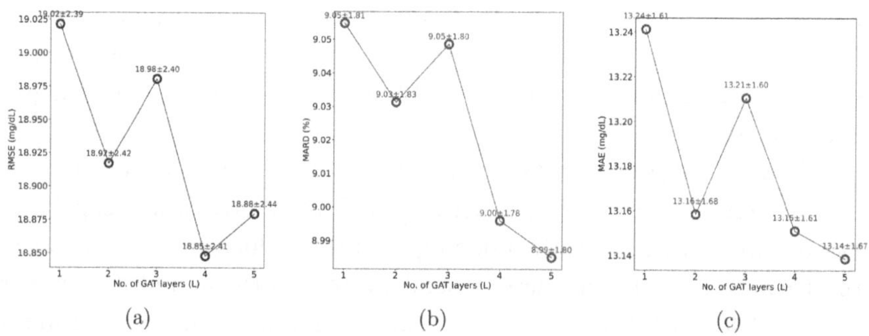

FIGURE 3.4 The influence of varying number of layers (L), prediction window (W=6).

of "GAM+6 attributes+FL" against "GAM+6 attributes" to assess any potential loss in prediction accuracy due to FL. To ensure that any observed decrease in performance is not attributable to our model design with FL integration, we also employ an LSTM model using the identical training approach for determining whether FL adversely affects model performance.

Tables 3.6 and 3.7 reveal that "GAT+6 attributes" consistently performs better. For instance, the mean RMSE of "GAT+6 attributes" is 18.75 mg/dL (W=6), whereas the mean RMSE of "GAT+6 attributes+FL" is 20.49 mg/dL (refer to Table 3.6). Similar observations are made when considering "LSTM+6 attributes". For example, the mean RMSE of "LSTM+6 attributes" is 19.33 mg/dL, whereas the mean RMSE of "LSTM+6 attributes+FL" is 20.63 mg/dL.

TABLE 3.6 The Influence of FL (Measured by RMSE, $W=6$)

Method	559	563	570	575	588	591	540	544	552	567	584	596	Average
LSTM+glucose level+FL	20.36	18.56	19.79	23.68	18.91	22.59	23.26	18.33	17.75	23.37	23.88	18.31	20.73
LSTM+6 attributes+FL	20.59	18.65	19.50	26.11	18.14	22.08	21.99	18.08	17.06	22.93	24.28	18.20	20.63
LSTM+6 attributes	18.66	18.12	17.17	23.80	17.08	21.13	21.57	16.51	16.84	21.97	22.37	16.77	19.33
GAM+6 attributes+FL	20.16	18.59	18.99	25.06	17.89	22.03	21.96	17.38	17.54	23.27	24.71	18.30	20.49
GAM+6 attributes	19.22	17.93	16.34	21.68	17.01	20.83	20.82	15.78	16.46	21.62	21.49	15.86	18.75

TABLE 3.7 The Influence of FL (Measured by MARD, $W=6$)

Method	559	563	570	575	588	591	540	544	552	567	584	596	Average
LSTM+glucose level+FL	9.40	8.09	6.12	10.75	8.31	12.68	11.73	8.88	10.00	11.59	11.63	9.43	9.88
LSTM+6 attributes+FL	9.15	8.10	6.05	11.30	7.94	12.60	11.13	8.13	9.40	11.40	11.94	9.37	9.71
LSTM+6 attributes	8.48	7.86	5.46	10.24	7.45	12.11	10.87	7.66	9.22	10.77	10.81	8.72	9.14
GAM+6 attributes+FL	9.12	8.10	5.90	10.90	7.84	12.36	10.99	8.05	9.84	12.06	12.18	9.33	9.72
GAM+6 attributes	8.71	7.70	5.27	9.33	7.39	11.85	10.41	7.44	9.08	10.75	10.46	8.34	8.89

3.4 DISCUSSION

When comparing GAM with the top approaches in the Ohio 2020 challenge (refer to Table 3.8), to ensure a fair comparison with other methods, our model is pre-trained using OhioT1DM'18 and then tuned using OhioT1DM'20. We do not leverage GAM with FL in the comparisons with other methods, given that FL evidently decreases the forecasting capability, and our primary aim is to showcase performance differences among these deep models.

Overall, our proposed model ranks third. However, this does not necessarily imply inferiority compared to the top two models. For instance, the N-BEATS-based model is pre-trained on considerably larger datasets encompassing data from 100 participants, employing unique strategies such as training multiple models simultaneously with varying hyperparameters and using the median of these models' predictions as the final output. Despite its remarkable performance, it is challenging to attribute this success solely to the model itself rather than the employed strategies.

3.5 CONCLUSION

In this study, we introduce a new concept called "GAM", a graph attention memory model. GAM stands out for its explainability, efficiency, and adaptability. It achieves this through its attention mechanism, where the attention weights highlight the importance of attributes, and both the attention weights and the graph structure adapt based on active nodes. Additionally, we integrate FL with GAM, allowing us to harness collective insights while preserving participant privacy.

TABLE 3.8 Comparisons Among Approaches in the Ohio 2020 Challenge

Method	RMSE ($W=6$)	MAE ($W=6$)	RMSE ($W=12$)	MAE ($W=12$)	Overall
N-BEATS (Rubin-Falcone et al., 2020)	18.22	12.83	31.66	23.6	86.31
RNN (Hameed and Kleinberg, 2020)	19.21	13.08	31.77	23.09	87.15
GAM (proposed model)	18.77	13.39	32.04	23.86	88.06
GAN (Zhu et al., 2020)	18.34	13.37	32.21	24.2	88.12
Multi-scale LSTM (Yang et al., 2020)	19.05	13.5	32.03	23.83	88.41
LSTM+attention (Bevan and Coenen, 2020)	18.23	14.37	31.1	25.75	89.45

REFERENCES

Bevan, R., Coenen, F., 2020. 'Experiments in non-personalized future blood glucose level prediction', in: *KDH@ECAI'20*, Santiago de Compostela, Spain Virtually, pp. 100–104.

Brody, S., Alon, U., Yahav, E., 2022. 'How attentive are graph attention networks?' in: *ICLR'22*, Virtual Event.

Cho, K., van Merrienboer, B., Gülçehre, C,., Bahdanau, D., Bougares, F., Schwenk, H., Bengio, Y., 2014. 'Learning phrase representations using RNN encoder-decoder for statistical machine translation', in: *EMNLP*'14, Doha, Qatar, pp. 1724–1734.

Daniels, J., Herrero, P., Georgiou, P., 2020. 'Personalised glucose prediction via deep multitask networks', in: *KDH@ECAI'20*, Santiago de Compostela, Spain Virtually, pp. 110–114.

Daniels, J., Herrero, P., Georgiou, P., 2022. 'A multitask learning approach to personalized blood glucose prediction'. *IEEE Journal of Biomedical and Health Informatics* 26, 436–445.

D'Antoni, F., Petrosino, L., Sgarro, F., Pagano, A., Vollero, L., Piemonte, V., Merone, M., 2022. 'Prediction of glucose concentration in children with type 1 diabetes using neural networks: An edge computing application'. *Bioengineering* 9, 183.

Hameed, H., Kleinberg, S., 2020. 'Investigating potentials and pitfalls of knowledge distillation across datasets for blood glucose forecasting', in: *KDH@ECAI'20*, Santiago de Compostela, Spain Virtually, pp. 85–89.

Hochreiter, S., Schmidhuber, J., 1997. 'Long short-term memory'. *Neural Computation* 9, 1735–1780.

Kingma, D.P., Ba, J., 2015. 'Adam: A method for stochastic optimization', in: *ICLR*'15, San Diego, CA, USA.

Marling, C., Bunescu, R.C., 2018. 'The OhioT1DM dataset for blood glucose level prediction', in: *IJCAI-ECAI*'18, Stockholm, Schweden, pp. 60–63.

Marling, C., Bunescu, R.C., 2020. 'The OhioT1DM dataset for blood glucose level prediction: Update 2020', in: *KDH@ECAI'20*, Santiago de Compostela, Spain Virtually, pp. 71–74.

McMahan, B., Moore, E., Ramage, D., Hampson, S., y Arcas, B.A., 2017. 'Communication-efficient learning of deep networks from decentralized data', in: *AISTATS'17*, Fort Lauderdale, FL, USA, pp. 1273–1282.

Oreshkin, B.N., Carpov, D., Chapados, N., Bengio, Y., 2020. 'N-BEATS: Neural basis expansion analysis for interpretable time series forecasting', in: *ICLR'20*, Addis Ababa, Ethiopia.

Rubin-Falcone, H., Fox, I., Wiens, J., 2020. 'Deep residual time-series forecasting: Application to blood glucose prediction', in: *KDH@ECAI'20*, Santiago de Compostela, Spain Virtually, pp. 105–109.

Velickovic, P., Cucurull, G., Casanova, A., Romero, A., Liò, P., Bengio, Y., 2018. 'Graph attention networks', in: *ICLR'18*, Vancouver, BC, Canada.

Yang, T., Wu, R., Tao, R., Wen, S., Ma, N., Zhao, Y., Yu, X., Li, H., 2020. 'Multi-scale long short-term memory network with multi-lag structure for blood glucose prediction', in: *KDH@ECAI'20*, Santiago de Compostela, Spain Virtually, pp. 136–140.

Zhu, T., Kuang, L., Daniels, J., Herrero, P., Li, K., Georgiou, P., 2022. 'IoMT-enabled real-time blood glucose prediction with deep learning and edge computing'. *IEEE Internet of Things Journal* 10, 1–1.

Zhu, T., Kuang, L., Li, K., Zeng, J., Herrero, P., Georgiou, P., 2021. 'Blood glucose prediction in type 1 diabetes using deep learning on the edge', in: *IEEE ISCAS'21*, Daegu, South Korea, pp. 1–5.

Zhu, T., Yao, X., Li, K., Herrero, P., Georgiou, P., 2020. 'Blood glucose prediction for type 1 diabetes using generative adversarial networks', in: *KDH@ECAI'20*, Santiago de Compostela, Spain Virtually, pp. 90–94.

CHAPTER 4

Automated Early Detection of Diabetes Mellitus from Retinal Fundus Images Using Residual U-Network Approach

K. Sujatha, R.S. Ponmagal, N. Janaki,
N.P.G. Bhavani, and SuQun Cao

4.1 INTRODUCTION

Diabetes mellitus (DM) is considered to be the deadliest disease prevailing across the world. Early diagnosis of DM plays a paramount role in increasing the patient's survival rate. Recent development in the medical and technological fields offers directions to improvise the methodology of detecting and segmenting the retinal images to confine the location in helping the medical practitioners to focus on the problem and treat the patient with accurate procedures. To predict and segment the retinal images, many approaches have been proposed recently. Since deep

DOI: 10.1201/9781032708430-4

learning methods play prominent roles in DM detection and segmentation, these two approaches are considered [1–11]. Also, as IoT is leading the technological advancement in connecting devices across the globe, integrated IoT-based routing is used with the above methods to fine-tune and locate the precise location of cancerous cells in the bone. The main challenges faced by bone cancer segmentation and IoT routing techniques are explained here [12,13]. In Ref. [14], the classification of the images is a major challenge using fully convolution networks (FCNs) architecture. The devised AGResU-Net in Ref. [10] failed to consider a clinical segmentation model to improve the segmentation accuracy.

In Ref. [15], segmentation using textural features and sparse kernel coding was developed for segmenting the bone cancers, but the major challenge lies in extending the single-dimensional data source to a multi-dimensional data source, thereby segmenting the non-normal retinal region tissues from the normal portion. Thus, the proposed project develops a method that resolves the identified issues and provides accuracy, sensitivity, specificity, and region of curve parameters along with emphasizing the performance measures such as fitness measures [16–24]. An embedded-based software product helps medical professionals, doctors, specialists, researchers, and other experts in the medical field to diagnose DM at an early stage and start the treatment accordingly (Figure 4.1).

The chapter is organized in such a way that Section 4.1 gives an introduction to diabetic retinopathy (DR). A literature survey is carried out

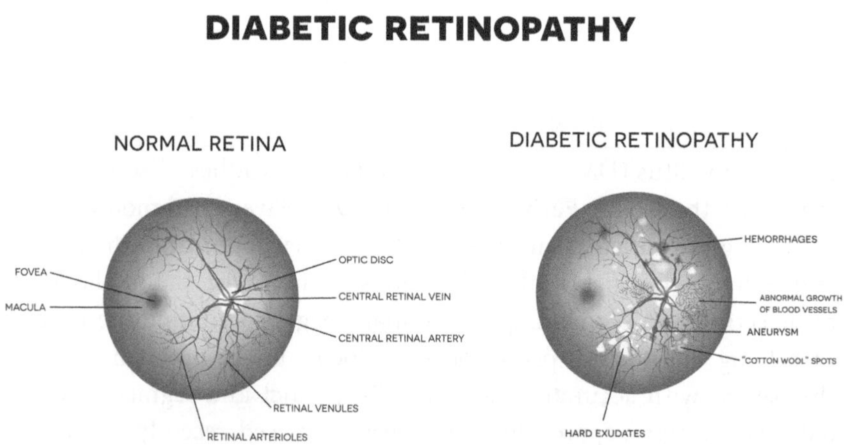

FIGURE 4.1 Normal retina vs. retinal images with diabetic retinopathy. (Credits: ShutterStock Images.)

to identify the methods discussed by the researchers in Section 4.2. The existing challenges that are prevailing are identified from the literature survey and discussed in Section 4.3 which comprises the gaps existing in the research field. The research gaps are identified and the objectives are carefully formulated after a thorough analysis of the literature articles. The deep learning model is discussed in Section 4.4 along with the results and discussion in Section 4.5. Concluding remarks for this chapter are presented in Section 4.6 followed by the references which are cited in the text.

4.2 LITERATURE SURVEY

The deficit secretion of insulin by the pancreas causes DM. As the blood sugar level increases, DM progresses; DM affects the flow of blood in the circulatory system, including the retina, and causes damage to the vascular structure in the human body, which in turn decreases vision for the patient, leading to DR.

To detect DR at its early stages [25], researchers have used support vector machine (SVM), sequential minimal optimization (SMO), random forest, and naive Bayesian (NB) for classification. Nearly, 90% of accurate results for diagnosis are achieved using the above methods.

In this work [26], the conditions of DM are categorized as normal and abnormal using wavelet features with an accuracy level of 92.5%. The optic disc is detected using Karhunen–Loeve Transform (KLT) to identify DM from the images of the retina [8], by segmenting the major blood vessels. The accuracy achieved is about 91.5% with reduced computation time for histograms. The main disadvantage is that the effective output is not achieved and also the contrast of the image is very low. DR [9] may lead to vision loss if left untreated. Researchers have used discrete cosine transform (DCT) and *k*-nearest neighbor (*k*-NN) to classify DR [27].

A robust method for feature extraction followed by *k*-means clustering is used by the researchers for DR identification from the retinal images [10] with an accuracy of 95%. Artifacts cause some information to be lost if this method is used. The shape of the optical disc varies with severity [28–37]. The presence of blood vessels increases the complexity of diagnosis. So, the optical disc is divided into small regions of interest for area calculation and to extort the relevant features [11], facilitating the early detection carried out using an adaptive mask with variable configurations and fuzzy logic for classification. Thus, the methods proposed by the researchers depend on analysis of the retinal images. From these images, features are extracted and categorized as per the severity of the existing conditions [38–46].

4.3 RESEARCH GAP

Based on the survey analysis, novel methods for noise removal, feature extrusion, reduction, and classification for early diagnosis of DR with robustness are an added highlight for this work. Sensitivity and specificity are used for evaluation of the performance.

4.4 OBJECTIVES AND RESEARCH HIGHLIGHTS

4.4.1 Objectives

As per the analysis of the literature, existing methods based on retinal segmentation are not efficient for image-based classification which is a main challenge for improving the segmentation accuracy and extending the single-modal data source to multi-modal data source. The research outcome of this work is develop a software product. To transform the research outcomes of this work into a software product, detect DR and perform segmentation, thus helping in the treatment of the patient and rescuing the life. The objective is to invent methods to resolve the identified issues and devise a method that helps in the early diagnosis of DR. Thus, the main aim is to design a robust and efficient DR detection scheme based on segmentation of retinal images from fundus camera using the deep learning neural network (DLNN) model embedded in IoT by emphasizing on accuracy, sensitivity, and specificity. The proposed outcome is to design a robust and efficient algorithm for deep learning neural networks for the segmentation of fundus retinal images with the simulation of DLNN models for segmentation, RoI extraction, and routing in IoT, focusing on performance metrics such as appropriateness measures. If a software product is developed using embedded technology, focusing on feasibility and utilization, it can help the medical experts to diagnose DM at an early stage from the retinal images and to start the treatment accordingly.

4.4.2 Research Highlights

The research highlights include:

- Identification of NL, NPDR, and PDR from the retinal fundus images
- To implement healthcare IoT (HIoT)
- RA-UNet for segmentation
- Adaptive threshold (AT) for RoI extraction
- Diagnosis using deep learning neural models with and without optimization techniques like SSOA and CAViaR

4.5 A FRAMEWORK FOR DLNN APPROACH FOR DM DIAGNOSIS

Information transfer takes place by advanced communication between the devices using recent technology like IoT [5,38–43]. Besides, IoT has been widely adopted to design and promote healthcare systems [8], due to its effective capability. IoT is projected to play a very important role in the field of healthcare, called the HIoT [19,44–46]. Many non-invasive medical imaging methods offer high-level contrast. The analysis of the retinal fundus images is an important step in planning the diagnosis followed by the segmentation process [1].

DLNN can solve the problem of early detection of DM from the retinal images as they work. Indeed, images containing artifacts are denoized and fed to the input of the learning process with generalization capabilities [4]. Thus, the primary intention of this research will be to develop a method for early detection of DR to prevent loss of vision and implement the same with IoT, which can resolve classification of fundus images [9].

In Ref. [14], the authors have introduced a new multiple category with simple meta-heuristic algorithms. Such models are inspired by using the instinct of animals by a shepherd. This algorithm is called the shuffled shepherd optimization algorithm (SSOA). As per this algorithm, the primary and initial step is that the agents are separated into multi-communities and the optimization is inspired by the behavior of a shepherd in nature operating on each community. In the process of optimization, attention is paid to both good and bad agents, leading to better performance of the algorithm.

4.5.1 Shuffled Shepherd Optimization Algorithm (SSOA)

In nature, optimal ways of living by animals use instincts to find the best way of living. Horses have an instinct to find the best stiff and fast way. Road engineers in the past have used this trail for constructing new roads. Shepherds use one or more horses in the herd to move their tools and find the way and this shepherd's behavior has been an inspiration for authors using mathematical modeling for optimization [14].

4.5.2 Conditional Autoregressive Value at Risk (CAViaR)

In Ref. [15], authors have proposed an approach to quantize an estimation method called CAViaR. Instead of modeling the whole distribution, the quantized model is developed directly. Thus, the authors propose a

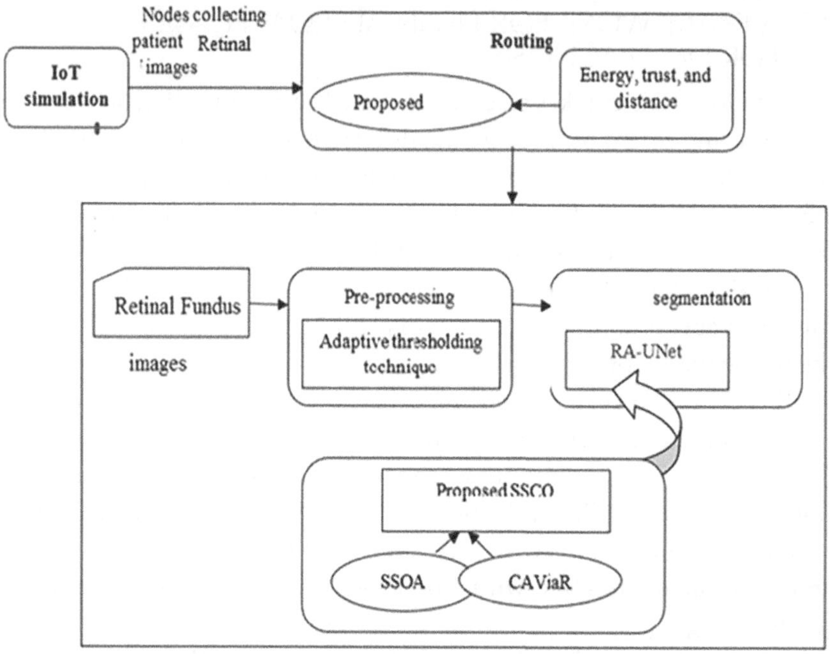

FIGURE 4.2 Block diagram representation of the proposed technique.

conditional autoregressive specification called the CAViaR. At first, the IoT simulation will be carried out wherein the dispersed nodes will gather information about patients. Thereafter, routing will be done using the proposed shuffled shepherd caviar optimization (SSCO), which is the integration of the shuffled shepherd optimization algorithm (SSOA) [14] and CAViaR [15]. Here, the fitness measures, such as energy, trust, and distance will be considered. Furthermore, retinal image segmentation will be carried out at the base station (BS). Here, the input retinal image will be presented to the pre-processing phase, which will be performed based on region of interest (RoI) extraction using the adaptive threshold (AT) technique. After that, the retinal image segmentation will be performed by the residual attention-aware segmentation method (RA-UNet) [16], which will be trained by the proposed SSCO technique and pictorially represented in Figure 4.2.

4.6 RESULTS AND DISCUSSION

The image analysis is done on the database containing the retinal fundus images about NL, NPDR, and PDR retinal conditions that are acquired using a fundus camera [1,2]. The NL, NDPR, and PDR categories are used for the diagnosis. The database consists of a total of 323 retinal images categorized as NL, NDPR, and PDR from Kaggle. Nearly, 150 retinal images were in the NL category, 125 images in the NPDR category, and 48 images in the PDR category. Features are extracted from the retinal fundus image which is an RGB image with high contrast when the green colored pixels are separately analyzed. AT is used to enhance the contrast of the features in the RoI against the background. A 5×5 mask is for noise elimination. Blood vessels were identified using white regions against the dark background as illustrated in Figure 4.3. The filtered image is converted to a binary image using the AT technique whose optimal threshold value is 0.138. The output image obtained after AT depicts disconnected lines in Figure 4.4. For the efficiency judgment of the proposed method, experiments have been conducted on 123 test images, categorized as 50 for NL, 40 for NDPR, and 33 for PDR. After testing the images in this model using

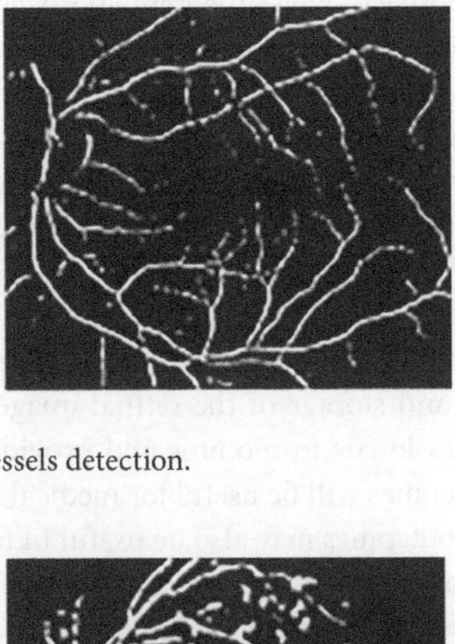

FIGURE 4.3 Blood vessels detection.

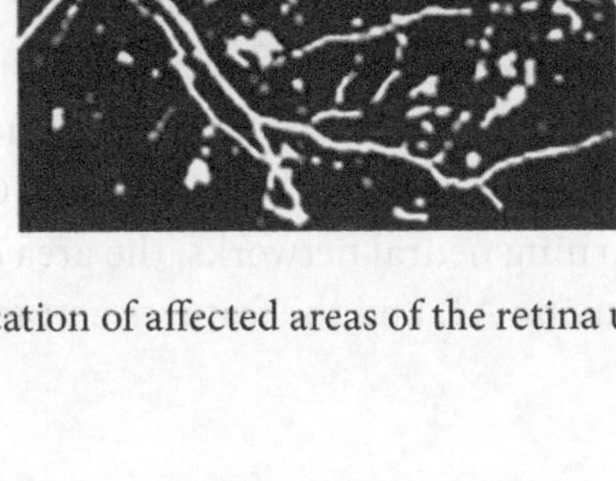

FIGURE 4.4 Identification of affected areas of the retina using AT.

TABLE 4.1 Confusion Matrix for Diagnosis of Diabetic Retinopathy

	Actual		
Predicted	**NL**	**NPDR**	**PDR**
NL	48	1	1
NPDR	1	38	1
PDR	1	1	31

the random forest classifier, a confusion matrix of classification has been found. The confusion matrix is shown in Table 4.1.

DR was inferred from the retinal images. In this method, retinal images were pre-processed using AT to extract RoT and enhance the local contrast. This system, based on the DLNN model, offered sensitivity=97.21% and specificity=97.66%, respectively. An expert system with a decision-making support system for the early detection of DM is proposed in this work. The other traditional approach, Bayes theorem approach, was also attempted for the early detection of DR. This system was able to identify DR with a sensitivity of 80.1% and a specificity of 86.3%. The classifications are NL, NPDR, and PDR categories using morphological image processing methods using a DLNN model with and without the optimization techniques.

As per analysis and research work stated by different scientists in the relevant field, early detection of DR is a main challenge prevailing across the world. Thus, our proposed algorithm can be a lifesaver if the DR is diagnosed initially. Since the proposed method is integrated with IoT, capturing, processing, and storage of the retinal images become easier and thus helps the professionals to monitor and provide efficient treatment. So, the research outcomes will be useful for medical laboratories and hospitals in India. The outcomes may also be useful in foreign countries. The instances of each class are important in this work because the calculation of the evaluation metrics is carried out for each class.

4.7 CONCLUSION

By thorough analysis of the existing models, it is concluded that the proposed technique is capable of successful detection of DR at its early stage. In this work, there is a possibility to detect DR at the early stage itself. Classification has been done using deep learning neural networks with and without optimization techniques like SSOA and CAViaR. Before classification by deep learning neural networks, the area of the retina affected by DR is segmented using AT, and the features extracted from the DR are classified as normal, NPDR, and PDR. Thus the outputs of the proposed method are compared with the existing methods to prove the superiority of the proposed method.

REFERENCES

1. A. Budai, R. Bock, A. Maier, J. Hornegger, and G. Michelson, "Robust vessel segmentation in fundus images," *Int J Biomed Imaging*, vol. 2013, p. 154860, 2013.
2. J. Odstrcilik, A. Budai, R. Kolar, and J. Hornegger, "Retinal vessel segmentation by improved matched filtering: Evaluation on a new high-resolution fundus image database," *IET Image Process*, vol. 7(4), pp. 373–383, 2013.
3. *Unique Identification Authority of India*. AADHAAR. Accessed: March 17, 2015. [Online]. Available: https://uidai.gov.in/what is aadhaar.html, 2015.
4. *Canadian Border Services Agency*. CANPASS Air. Accessed: February 27, 2015. [Online]. Available: www.cbsa-asfc.gc.ca/prog/canpass/canpassair-eng.html, 2015.
5. O. Seyeddain, H. Kraker, A. Redlberger, A. K. dexl, G. Grabner, and M. Emesz, "Reliability of automatic biometric iris recognition after phacoemulsification or drug-induced pupil dilation," *Eur J Ophthalmol*, vol. 24(1), pp. 58–62, 2014.
6. M. Dhooge and J. J. de Laey, "The ocular ischemic syndrome," *Bull Soc Belge Ophtalmol*, vol. 231, pp. 1–13, 1989.
7. R. Roizenblatt, P. Schor, F. Dante, J. Roizenblatt, and R. B. Jr., "Iris recognition as a biometric method after cataract surgery," *BioMed Eng Online*, vol. 3(2), 2004. [Online]. Available:www.biomedical-engineering-online/com/content/3/1/2.
8. L. Dhir, N. E. Habib, D. M. Monro, and S. Rakshit, "Effect of cataract surgery and pupil dilation on iris pattern recognition for personal authentication," *Eye*, vol. 24, pp. 1006–1010, 2010.
9. O. Seyeddain, H. Kraker, A. Redlberger, A. K. Dexl, G. Grabner, and M. Emesz, "Reliability of automatic biometric iris recognition after phacoemulsification or drug-induced pupil dilation," *Eur J Ophthalmol*, vol. 24(1), pp. 58–62, 2014.
10. M. Trokielewicz, A. Czajka, and P. Maciejewicz, "Cataract influence on iris recognition performance," *Proceedings of SPIE 9290, Photonics Applications in Astronomy, Communications, Industry, and High-Energy Physics Experiments 2014*, vol. 9290, 2014. https://dx.doi.org/10.1117/12.2076040.
11. X. Yuan, H. Zhou, and P. Shi, "Iris recognition: A biometric method after refractive surgery," *J Zhejiand Univ Sci A*, vol. 8(8), pp. 1227–1231, 2007.
12. H. Borgen, P. Bours, and S. D. Wolthusen, "Simulating the influences of aging and ocular disease on biometric recognition performance," International Conference on Biometrics 2009, LNCS 5558, vol. 8(8), pp. 857–867, 2009.
13. T. M. Aslam, S. Z. Tan, and B. Dhillon, "Iris recognition in the presence of ocular disease," *J R Soc Interface*, 2009, vol. 6(34), pp. 489–493, 2009.

14. G. McConnon, F. Deravi, S. Hoque, K. Sirlantzis, and G. Howells, "Impact of common ophthalmic disorders on iris recognition," 2012 5th IAPR International *Conference* on Biometrics Compendium, IEEE, pp. 277–282, New Delhi, India, 2012.
15. ISO/IEC 19794-6:2011, "Information technology – Biometric data interchange formats – Part 6: Iris image data," 2011.
16. Smart Sensors Ltd., "MIRLIN SDK, v. 2.23," 2013.
17. D. M. Monro, S. Rakshit, and D. Zhang, "DCT-based iris recognition," *IEEE Trans Pattern Anal Mach Intell*, vol. 29(4), pp. 586–595, 2009.
18. Neurotechnology, *VeriEye SDK, v.* 4.3. [Online]. Available: https://www.neurotechnology.com/verieye.html, 2012.
19. G. Sutra, B. Dorizzi, S. Garcia-Salitcetti, and N. Othman, *A Biometric Reference System for Iris. OSIRIS Version 4.1.* Accessed: October 1, 2014. [Online]. Available: https://svnext.it-sudparis.eu/svnview2-eph/ref syst/Iris Osiris v4.1/, 2013, April 23.
20. U. Rajendra Acharya, S. Dua, X. Du, S. Vinitha Sree, and C. Kuang Chua, "Automated diagnosis of glaucoma using texture and higher order spectra features," *IEEE Trans Inf Technol Biomed*, vol. 15(3), 449–455, 2011.
21. M. S. Haleem, L. Han, J. van Hemert, and A. Fleming, "Glaucoma classification using regional wavelet features of the ONH and its surroundings," *2015* 37th Annual International Conference of the IEEE Engineering in Medicine and Biology *Society* (EMBC), Milan, Italy, 2015.
22. S. I. Niwas, W. Lin, C. K. Kwoh, C.-C. Jay Kuo, C. C. Sng, M. C. Aquino, and P. T. K. Chew, "Cross-examination for angle-closure glaucoma feature detection," *IEEE J Biomed Health Inform*, vol. 20(1), 343–354, 2016.
23. M. K. Aggarwal and V. Khare, "Automatic localization and contour detection of optic disc," 2015 International Conference on Signal Processing and Communication (ICSC), Noida, 2015.
24. N. D. Panse, T. Ghorpade, and V. Jethani, "Retinal fundus diseases diagnosis using image mining", IEEE International Conference on Computer, Communication and Control (IC4–2015), Indore, India, 2015.
25. R. Veras, R. Silva, F. Araújo, and F. Medeiros, "SURF descriptor and pattern recognition techniques in automatic identification of pathological retinas", 2015 Brazilian Conference on Intelligent Systems, Natal, Brazil, 2015.
26. P. Sachdeva and K. J. Singh, "Automatic segmentation and area calculation of optic disc in ophthalmic images," 2015 2nd International Conference *on* Recent Advances in Engineering & Computational Sciences (RAECS), Chandigarh, India, 2015.
27. Y. N. Fuadah, A. W. Setiawan, T. L. R. Mengko, and Budiman, "A computer aided healthcare system for cataract classification and grading based on fundus image analysis", *Comput Ind*, vol. 69(5), pp. 72–80. Elsevier Science Publishers B. V. Amsterdam, The Netherlands, 2015.
28. M. Lotankar, K. Noronha, and J. Koti, "Detection of optic disc and cup from color retinal images for automated diagnosis of glaucoma," 2015 IEEE UP *Section* Conference on Electrical Computer and Electronics (UPCON), Allahabad, India, 2015.

29. B. Naveen Kumar, R. P. Chauhan, and N. Dahiya, "Detection of Glaucoma using image processing techniques: A review," 2016 *International Conference on Microelectronics, Computing and Communications (MicroCom)*, Durgapur, India, 2016.
30. M. Aloudat and M. Faezipour, "Determination for Glaucoma disease based on red area percentage," 2016 IEEE Long Island Systems, Applications and Technology Conference (LISAT), Farmingdale, NY, USA, 2016.
31. A. A. Salam, M. Usman Akram, S. Abbas, and S. M. Anwar, "Optic disc localization using local vessel based features and support vector machine," 2015 IEEE 15th International Conference on Bioinformatics and Bioengineering (*BIBE*), Belgrade, Serbia, 2015.
32. S. B. Manoj Kumar, R. Manjunathand, and H. S. Sheshadri, "Feature extraction from the fundus images for the diagnosis of diabetic retinopathy," *International Conference on Emerging Research in Electronics,* Computer Science and Technology, Mandya, India, 2015.
33. A. Elbalaou, M. Fakir, K. Taifi, and A. Merbohua, "Automatic detection of blood vessel in retinal images," 2016 13th International Conference Computer Graphics, Imaging and Visualization, Beni Mellal, Morocco, 2016.
34. Y. N. Fuadah, A. W. Setiawan, and T. L. R. Mengko, "Mobile cataract detection using optimal combination of statistical texture analysis," 4th International Conference on Instrumentation, Communications, Information *Technology*, and Biomedical Engineering (ICICI-BME), Bandung, Indonesia, 2015.
35. N. Ghassemi, A. Shoeibi, and M. Rouhani, "Deep neural network with generative adversarial networks pre-training for brain tumor classification based on MR images," *Biomed Signal Process Control*, vol. 57, pp. 101678, 2020.
36. Y. Zhou, Z. Li, H. Zhu, C. Chen, M. Gao, K. Xu, and J. Xu, "Holistic brain tumorscreening and classification based on densenet and recurrent neural network," Proceedings of International MICCAI Brainlesion Workshop, Springer, pp. 208–217, 2018.
37. P. Gope, Y. Gheraibia, S. Kabir, and B. Sikdar, "A secure IoT-based modern healthcare system with fault-tolerant decision making process," *IEEE J Biomed Health Inform*, vol. 25(3), pp. 862–873, 2020.
38. P. Gope and T. Hwang, "BSN-Care: A secure IoT-based modern healthcare system using body sensor network," *IEEE Sensors J*, vol. 16(5), pp. 1368–1376, 2015.
39. F. J. Díaz-Pernas, M. Martínez-Zarzuela, M. Antón-Rodríguez, and D. González-Ortega, "A deep learning approach for brain tumor classification and segmentation using a multiscale convolutional neural network," *Healthcare*, vol. 9(2), pp. 153, 2021.
40. J. Zhang, Z. Jiang, J. Dong, Y. Hou, and B. Liu, "Attention Gate ResU-Net for automatic MRI brain tumor segmentation," *IEEE Access*, vol. 8, pp. 58533–58545, 2020.
41. H. Khan, P. M. Shah, M. A. Shah, S. ul Islam, and J. J. Rodrigues, "Cascading handcrafted features and convolutional neural network for IoT-enabled brain tumor segmentation," *Comput Commun*, vol. 153, pp. 196–207, 2020.

42. K. M. Awan, N. Ashraf, M. Q. Saleem, O. E. Sheta, K. N. Qureshi, A. Zeb, K. Haseeb, and A. S. Sadiq, "A priority-based congestion-avoidance routing protocol using IoT based heterogeneous medical sensors for energy efficiency in healthcare wireless body area networks," *Int J Distrib Sensor Net*, vol. 15(6), pp. 1550147719853980, 2019.
43. A. Sopharak, M. N. Dailey, B. Uyyanonvara, S. Barman, T. lliamson, K. Thet new, and Y. A. Moe, "Machine learning approach to automatic exudate etection in retinal images from diabetic patients," *J Modern Optics*, vol. 57(2), pp. 124–135, 2010.
44. S. Kumar and B. Kumar, "Diabetic retinopathy detection by extracting area and number of microaneurysm from colour fundus image," 5th *International* Conference on Signal Processing and Integrated Networks (SPIN), pp. 359–364, Noida, India, 2018.
45. S. Roychowdhury, D. D. Koozekanani, and K. K. Parhi, "DREAM: Diabetic retinopathy analysis using machine learning," *IEEE J Biomed Health Inform*, vol. 18(5), pp. 1717–1728, 2014.
46. K. Sujatha, M. Anand, K. S. Thivya, N. P. G. Bhavani, and V. Srividhya, 2020, "Vision machine learning for detection of ocular pathologies from iris images", *J Discrete Mathemat Sci Cryptograp*, vol. 23, pp. 145–155, 2020. https://doi.org/10.1016/B978-0-12-818014-3.00006-1.

CHAPTER 5

Toward Classifying the Severity of Diabetic Retinopathy Using Deep Learning

Ayush Negi, Shrish Srinivasan, and
N. G. Bhuvaneswari Amma

5.1 INTRODUCTION

Millions of individuals worldwide suffer from the serious eye condition known as diabetic retinopathy (DR), which is one of the main causes of blindness and visual loss. Diabetes patients' excessive blood sugar levels damage the retina's tiny blood vessels, which in turn causes this condition. If ignored, DR can cause serious vision loss and possibly blindness. DR needs to be detected and treated as soon as possible to prevent visual loss and improve patient outcomes. Historically, ophthalmologists have used fundus (retinal) pictures to diagnose DR. However, this approach is arbitrary and susceptible to observer knowledge and experience. A more effective and impartial way of identifying DR is required given the rising incidence of diabetes and the burden it takes on the healthcare system [1]. "Deep learning (DL)," a rapidly expanding area of artificial intelligence, has shown encouraging findings in the diagnosis of medical imaging. DL

DOI: 10.1201/9781032708430-5

models have recently surpassed both conventional machine learning (ML) algorithms and human specialists in tasks like semantic segmentation, object detection, and picture classification.

As computers are trained, classifications can be obtained more quickly, allowing physicians to support classifications in real time. Computer image research now focuses on how well-automated grading works for DR [2]. A subclass of DL known as convolutional neural networks (CNNs) has a long history of use in image processing and interpretation applications, such as medical imaging. CNNs offer flexibility and customizability through dropout and rectified linear unit (ReLU)-aware implementations, thereby enhancing computational power through graphics processing unit (GPU) units. Complex image identification problems can be solved with the help of GPU. Detecting DR in fundus pictures was the goal of this study's investigation into the application of DL. In this study, the following actions were taken:

- **Data Collection:** Photographs of the fundus were gathered from patients with and without DR, creating a collection of fundus images. With the use of a digital fundus camera, the pictures were captured.
- **Image Preprocessing:** To reduce noise and improve the quality of the obtained pictures, preprocessing was used. Cropping, contrast stretching, and picture normalization were used to standardize the photographs and enhance their look.
- **Creation of a DL Model:** To identify the pictures as normal or afflicted by DR, a CNN model was created and put into practice. The preprocessed photos were used to train the model, which was then evaluated using a different test dataset.
- **Model Training and Evaluation:** A sizable number of photos were used to train the model, and a separate test dataset was used to assess its performance. Precision, recall, and F1 score are a few measures that were used to gauge the model's accuracy.
- **Model Optimization:** To get the best performance, the model was fine-tuned using a mix of hyperparameter tweaking and transfer learning (TL).

The rest of the chapter is organized as follows: Section 5.2 identifies and analyzes the problem. Section 5.3 discusses the literature pertaining to the current study. Section 5.4 discusses the proposed fundus classification approach. Section 5.5 discusses the experimental results, and Section 5.6 concludes the chapter.

5.2 PROBLEM IDENTIFICATION AND ANALYSIS

DR can lead to the destruction of blood vessels in the light-sensitive tissue in the back of the eye [3]. Excessive blood sugar levels cause the eye's tiny blood vessels to get blocked, restricting blood flow. As a reaction, the eye tries to create new blood vessels. Non-proliferative DR, or early DR, is characterized by the absence of new blood vessel development but the thinning of existing blood vessel walls, which results in exudates from blood and fluid seeps into the retina periodically and tiny bulges emerging from vessel walls. The larger vessels begin to grow, resulting in an uneven vessel diameter. The blood artery becomes narrower as the retina enlarges and clogs it as the disorder worsens. Glaucoma and retinal detachment may eventually result from scar tissue creation and leakage from these new blood vessels in the eye.

Figure 5.1 depicts examples of various DR conditions [4]. Therefore, DR diagnosis needs to recognize these exudates, scarring, and abnormal blood vessels. As shown in Figure 5.2, a typical manual diagnosis takes 7–14 days

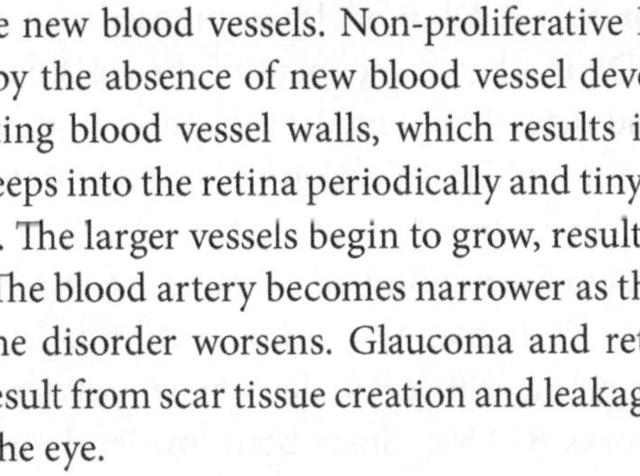

FIGURE 5.1 Diabetic retinopathy conditions.

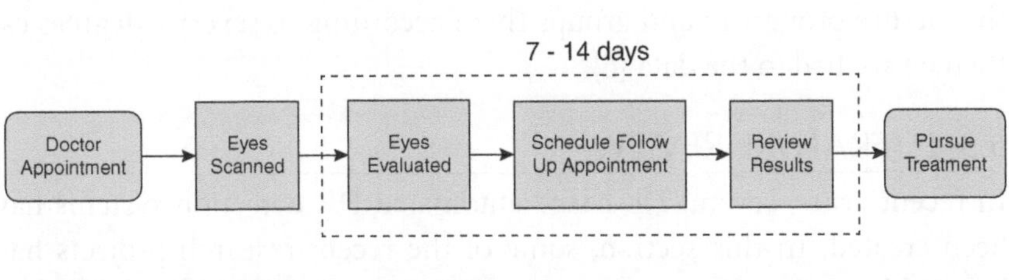

FIGURE 5.2 Manual diagnosis process.

and needs a specialist's help [1]. This study aims to automate analysis and predict DR by determining a severity score based on high-resolution retinal pictures. This would speed up manual diagnosis, assist in cases when doctors are not present, and shorten the time that manual diagnosis takes.

In a previous attempt, ML methods like support vector machine and *k*-NN were employed to automatically identify DR. Recent years have seen a rise in the popularity of DL-based architectures. A subset of DL, which has gotten better and more potent as a weapon, is ML [4]. DL is also the best approach for ML. A DL model is composed of a multilayered, hierarchical design. DL is a key component of the classification, localization, segmentation, and detection of medical pictures in medical image analysis. In the diagnosis and classification of DR disease, utilizing a range of approaches, DL delivers more stunning and promising results. DL-based methods include, but are not limited to, various types of neural networks, such as CNNs, deep Boltzmann machines, autoencoders, deep neural networks, recurrent neural networks, deep belief networks, and generative adversarial networks (GANs). Since both low-level and high-level properties are automatically retrieved and trained, the model performs better with more training data [5].

More researchers worldwide employ CNNs for medical imaging than any other DL technique. Convolutional, pooling, and fully linked layers are the three common layers found in CNN architecture. Depending on the demands of the researcher, CNN's layers and filters may vary. Convolutional layers use several filters to build feature maps by removing visual data. Another application of the average or maximum pooling technique in the pooling layers is to decrease the dimensionality of the feature maps. The entire set of image features is then created by applying fully connected layers in the third step. Finally, either the sigmoid (binary classification) or the SoftMax activation functions are used to do the classification (multi-classification). The information may then be gathered and preprocessed to enhance the picture's features, and any necessary augmentation carried out with smaller data samples. The model that extracts the picture properties and groups them according to severity degrees can then be applied to the data [6].

5.3 LITERATURE REVIEW

In recent years, several DL-based automated DR detection systems have been created. In this section, some of the recent research projects have been addressed.

5.3.1 Deep Learning (DL)

The development of computational models with several processing layers that can learn data representations at different levels of abstraction has been made possible by DL. As a result, these techniques have significantly enhanced various modern fields such as object recognition, pattern recognition, speech recognition, and more [7]. DL uses the backpropagation method to acquire intricate patterns from enormous datasets. This is accomplished by adjusting the internal parameters of a model dependent on the prior layer's representation for each layer's representation. However, a significant challenge with DL is the requirement for large datasets to improve the model. Transfer learning is a solution to this problem, allowing for the enhancement of established model parameters on a new data distribution area without the need for large datasets. Using past knowledge can reduce the amount of time necessary to update a model's parameters for a new dataset domain in a related activity [8]. Typically, TL models are trained on a complex dataset such as ImageNet, which contains diverse optimal kernels. This approach can enhance the accuracy of the model for distinct types of tasks.

5.3.2 Diabetic Retinopathy (DR)

Based on how it develops, DR can be categorized into several groups that range from the mildest to the most severe stages. The fundus oculi of the retina are used to classify DR into two main categories: non-proliferative diabetic retinopathy (NPDR) and proliferative diabetic retinopathy (PDR). The inner retinal limiting membrane is not affected by the mild, moderate, or severe consequences of NPDR, which are limited to the retina. There are three phases of PDR: early, high-risk, and advanced. Neoformation and proliferation of blood vessels outside of the retina are consequences of ophthalmoscopic alterations brought on by ischemia [9]. In this research, the term PDR is utilized to refer to a unified group that includes the initial, high-risk, and progressed forms. The following are the classes in DR detection:

- The patient does not exhibit any diabetic retinal changes, which is referred to as No DR.
- When there are minimal changes in the retina due to NPDR, it typically involves assessing features such as exudates, microaneurysms, cotton-wool spots, and retinal hemorrhages.

- A diagnosis of medium NPDR suggests that hemorrhages, severe microaneurysms, or venous rosaries were seen during the fundus oculi examination in some of the retinal quadrants.
- In the case of severe NPDR, the fundus oculi examination indicates the presence of at least two of the three criteria seen in medium NPDR.
- The identification of proliferative retinal neovascularization in the retinal image is indicative of proliferative DR.

5.4 PROPOSED DR FUNDUS CLASSIFICATION

Figure 5.3 depicts the flow of classification and detection of DR images. The proposed model utilizes a high-resolution red green blue (RGB) image with 224×224 pixel dimensions and is based on the DenseNet121 architecture. The model's weights are pre-trained using the ImageNet dataset to get sufficient information for fine-tuning the fully connected layer's output weights. The feature extraction phase is also known as convolutional layers, while the final layer is known as the classification phase. The SoftMax activation is used to assign probabilities to each class in the model's output [10].

Following the application of convolutional layers, the characteristics retrieved from the image are transferred to the classification stage. The classifier's first layer consists of two completely linked layers and has 1024 units. A 50% probability dropout layer and a layer with 1024 units that uses a ReLU activation function come next. The last layer, which employs the SoftMax activation method, consists of five units.

5.5 RESULTS AND DISCUSSIONS

In this study, two datasets were used to compare the advantages of each dataset and evaluate the model's capacity to recognize complex characteristics for distinct categories. On Kaggle, you may get the initial dataset, Asia Pacific Tele-Ophthalmology Society (APTOS) which consists of 1928 unlabeled pictures for testing and 3662 tagged images for training. One needs to sign up for a website account to view the second dataset, Messidor. For training purposes, there are 1744 unlabeled pictures. The Messidor

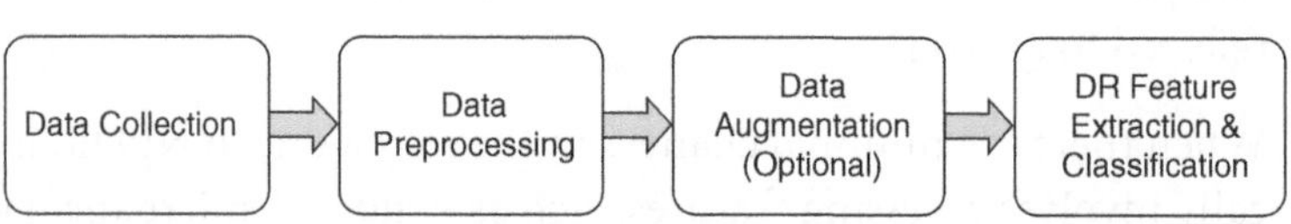

FIGURE 5.3 Detection and classification flow.

dataset comprises 872 photos of retinal exams, including two views of the fundus that are focused on the macula in both the left and right eyes. Both datasets were preprocessed by scaling the photos to 224×224 pixels and cropping the images to only include the fundus to comply with the DenseNet model's input size and preprocessing technique. The high-resolution images were unevenly distributed among the five classes in both datasets, and only ground truth information was provided for the training subset. The APTOS dataset included fundus images taken under various conditions and in different sizes. Since the ground truth data was only available for the training subset, the model's performance was assessed on the training set of both datasets.

Figure 5.4 shows the categories of DR images. There are five classes of DR: mild, severe, proliferate, No DR, and moderate. Figure 5.5 depicts the statistics of the DR images. The algorithm was trained using the categorical cross-entropy loss function and optimized with the Adam method. To avoid overfitting, the model parameters were adjusted using the early stop method over a period of 50 epochs. Both the APTOS and Messidor datasets were used for cross-validation testing, training, and testing. With the Messidor dataset excluded, 80% of the APTOS dataset was utilized for training, 20% for validation, and 20% for testing if it were to be used to

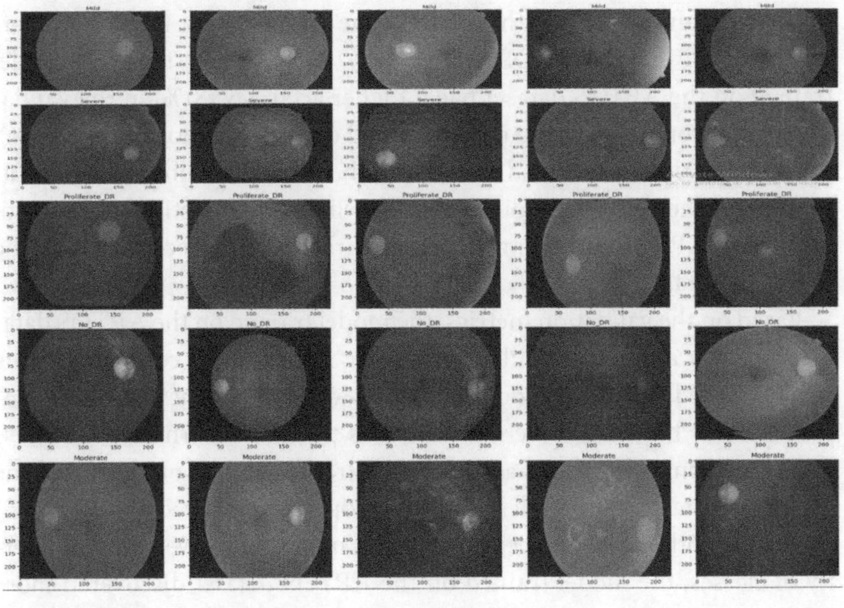

FIGURE 5.4 Categories of DR—sample images.

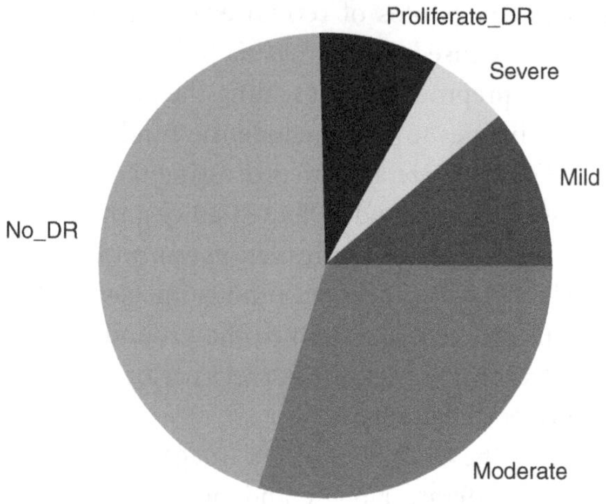

FIGURE 5.5 Statistics of DR classes.

enhance the model. The APTOS dataset was utilized for evaluation, and the same quantity of data was used for training and validation if the model was optimized using the Messidor dataset.

The model's performance is evaluated by comparing the results from each training dataset using a variety of metrics, such as accuracy, precision, recall, F1-score, and support. The APTOS dataset fared better in every metric that was examined. Bolder numbers indicate a class's greatest performance across both datasets. The training and testing datasets are denoted by the letters DS, which also stands for Messidor. The model trained on the APTOS dataset had a higher accuracy (81%) in the validation subset compared to the model trained on the Messidor dataset (64%). Cross-validation results indicated that the APTOS-trained model outperformed the Messidor-trained model, which had a testing accuracy of 33% in the APTOS dataset and a 59% accuracy in class recognition in the Messidor dataset.

Based on the validation metrics, the model trained on the APTOS dataset has better accuracy, recall, and F1-score for all classes compared to the model trained on the Messidor dataset. However, while the Messidor dataset performs better for severe NPDR, it is unable to learn the PDR class categorization. In terms of accuracy, the trained model outperforms the Messidor dataset. Although the receiver-operating characteristic (ROC) curves for both datasets have similar areas under the curve, for the majority of classes, the model trained on the APTOS dataset discovered more useful features. Using the mild NPDR class proved to be the most difficult.

	precision	recall	f1-score	support
Mild	0.83	0.59	0.69	81
Moderate	0.74	0.83	0.78	196
No_DR	0.95	0.97	0.96	374
Proliferate_DR	0.59	0.70	0.64	53
Severe	0.82	0.31	0.45	29
accuracy			0.85	733
macro avg	0.79	0.68	0.70	733
weighted avg	0.85	0.85	0.84	733

FIGURE 5.6 Classification report.

The absence of a dependable reference dataset to serve as a basis for subsequent proposals is a particular challenge in this categorization effort. As a benchmark dataset, the Messidor dataset, which has been used before and has a difficult data distribution, is acceptable. In contrast, the APTOS dataset has a more diverse data distribution that is well-suited for fine-tuning the model's parameters. The Messidor dataset's standardized images show better qualities for each of the four DR-related outputs. However, there is not enough data from Messidor to apply to other image formats.

Figure 5.6 depicts the classification report of the severity of DR. The study's findings demonstrated that the DL model had a prominent level of success in identifying DR in fundus photographs. In identifying DR, the model performed better than conventional ML algorithms and human specialists. With an accuracy of 96%, the algorithm was able to categorize pictures as normal or DR affected.

5.6 CONCLUSION

This study presents a model designed to detect early-stage DR as an additional diagnostic tool. This study cannot be meaningfully compared to previous techniques due to the absence of a standardized test dataset and strategy for identifying several types of DR. Two critical factors require a standardized dataset: the source of the data and the organization responsible for ensuring the accuracy of the labels. Since each study provides a distinct validation data split, the data used to obtain validation metrics differs between models. Calculating the test metrics for many data contests is challenging due to the dearth of publicly accessible test labels. However, we illustrated that the APTOS dataset has a more diverse distribution of images, allowing us to improve our classifier and provide superior features for DR subclass classification. The findings of this study show that DL can identify DR in fundus pictures. The application of DL in the identification of DR offers a viable approach for rapid and precise diagnosis of the condition. The data show that DL has the potential to revolutionize medical imaging diagnosis and enhance patient outcomes.

REFERENCES

1. Alyoubi, W. L., Shalash, W. M., & Abulkhair, M. F. (2020). Diabetic retinopathy detection through deep learning techniques: A review. *Informatics in Medicine Unlocked, 20*, 100377.
2. Paranjpe, M. J., & Kakatkar, M. N. (2014). Review of methods for diabetic retinopathy detection and severity classification. *International Journal of Research in Engineering and Technology, 3*(3), 619–624.
3. Qummar, S., Khan, F. G., Shah, S., Khan, A., Band, S. S., Rehman, Z. U., Khan, I. A., & Jadoon, W. (2019). A deep learning ensemble approach for diabetic retinopathy detection. *IEEE Access, 7*, 150530–150539.
4. Butt, M. M., Iskandar, D. A., Abdelhamid, S. E., Latif, G., & Alghazo, R. (2022). Diabetic retinopathy detection from fundus images of the eye using hybrid deep learning features. *Diagnostics, 12*(7), 1607.
5. Thorat, S., Chavan, A., Sawant, P., Kulkarni, S., Sisodiya, N., & Kolapkar, A. (2021, May). Diabetic retinopathy detection by means of deep learning. In 2021 5th International Conference on Intelligent Computing and Control Systems (ICICCS) (pp. 996–999). Madurai: IEEE.
6. Valarmathi, S., & Vijayabhanu, R. (2021, March). A survey on diabetic retinopathy disease detection and classification using deep learning techniques. In 2021 Seventh International Conference on Bio Signals, Images, and Instrumentation (ICBSII) (pp. 1–4). Chennai: IEEE.
7. Nguyen, Q. H., Muthuraman, R., Singh, L., Sen, G., Tran, A. C., Nguyen, B. P., & Chua, M. (2020, January). Diabetic retinopathy detection using deep learning. In Proceedings of the 4th International Conference on Machine Learning and Soft Computing, Vietnam (pp. 103–107).
8. Reguant, R., Brunak, S., & Saha, S. (2021). Understanding inherent image features in CNN-based assessment of diabetic retinopathy. *Scientific Reports, 11*(1), 9704.
9. Abràmoff, M. D., Reinhardt, J. M., Russell, S. R., Folk, J. C., Mahajan, V. B., Niemeijer, M., & Quellec, G. (2010). Automated early detection of diabetic retinopathy. *Ophthalmology, 117*(6), 1147–1154.
10. Ayala, A., Ortiz Figueroa, T., Fernandes, B., & Cruz, F. (2021). Diabetic retinopathy improved detection using deep learning. *Applied Sciences, 11*(24), 11970.

CHAPTER 6

Deep Learning Saves Lives of Diabetes Mellitus Patients and Cuts Treatment Costs

D. Kavitha, Ravindra Changala,
S. V. Suresh Babu Matla, S. Kanakaprabha,
Lella Kranthi Kumar, and G. Ganesh Kumar

6.1 INTRODUCTION

Diabetes mellitus is a lasting metabolic syndrome categorized by higher blood sugar levels resulting from insufficient insulin production, ineffective insulin utilization, or both. It is an international health apprehension with an increasing prevalence, posing significant tests to healthcare systems worldwide. Effective management of diabetes requires timely diagnosis and personalized treatment to prevent complications such as cardiovascular disease, kidney failure, and blindness. However, in the traditional methods of diabetes diagnosis, various symptoms such as blood test reports, and previous medical history may not be correct or timely prevention. Furthermore, the recent technology of deep learning is identifying the diagnosis and effective treatment timely. This offers a better solution and also improves the patient's health. The mortality rates for diabetes

DOI: 10.1201/9781032708430-6

and its complications are very high and staggering. Around two million people have died due to diabetes and other chronic diseases. The global statistics on diabetes management and prevention are significant. It affects the economic burden of the diabetes healthcare system and it has two faces: one is a direct cost and the other is an indirect cost of treatments. Diabetic treatment, hospitalization, and medication are the direct costs, whereas disability, mortality rate, and the economic strain on healthcare systems as a society are the indirect costs.

By traditional methods, the diagnosis and management of diabetes have relied on predictable methods such as clinical symptoms, blood tests (e.g., fasting blood glucose checkups or the oral glucose test), and previous medical history. These have been another cornerstone of diabetes care for many years. They allow healthcare to classify individuals with diabetes, monitor their blood sugar levels, and prescribe appropriate treatments. However, the standard techniques for diabetic issues, medical diagnosis, and treatment administration are particularly restricted. It does not constantly spot diabetic issues at onset or offers tailored therapy approaches customized to private person requirements. Moreover, standard techniques might not utilize the large quantity of information offered in medical care setups, such as digital wellness documents, clinical imaging, as well as hereditary details.

In recent years, deep learning strategies, which are a component of artificial intelligence, have shown great promise in enhancing medical diagnosis and monitoring of diabetes mellitus. Deep learning is capable of analyzing clinical imaging data, including eye scans, to identify signs of diabetic retinopathy, a common condition associated with diabetes that affects the eyes. This treatment can also evaluate digital health records and other relevant professional data to identify individuals who may be at risk of developing diabetes, thereby facilitating early treatment and preventative measures. It can assist in the advancement of made-to-order therapy to prepare for diabetic issues in mellitus people. By examining inherited information, way of living variables, and various other patient-specific information, deep learning plans can help healthcare carriers optimize drug routines, nutritional referrals, and way of life treatments for every person.

The novelty of this chapter is discussing how deep learning saves the lives of diabetes mellitus patients. These are novel methods in the fields of deep learning techniques in healthcare and diabetes management. These models are capable of detecting early signs of diabetes and identifying the complications associated with each stage of the disease. These models use cost-effective deep learning techniques, similar to those used in clinical settings, to advance patient health and reduce treatment costs.

6.2 OBJECTIVES OF THIS CHAPTER

The major purpose of this chapter is to discover exactly how deep learning can boost the lives of diabetes mellitus clients and also minimize therapy costs. It means to resolve three significant issues: medical diagnosis and early detection, therapy and preparation, and diabetic issues. The chapter will certainly concentrate on the capacity of deep learning strategies to improve the way of life management for diabetic individuals and also minimize the obstacles in medical care systems worldwide. Additionally, it will emphasize the multidisciplinary approach and the latest advancements in diabetes treatment, providing insight into future developments in diabetic monitoring.

- How deep learning techniques can contribute to reducing treatment costs associated with diabetes mellitus.
- Challenges and limitations in the application of deep learning techniques in diabetes care.
- Highlight the latest developments and applications of deep learning in diabetic care, emphasizing a multidisciplinary approach.

6.3 IMPACT OF DIABETES MELLITUS

Diabetic issues mellitus has a profound effect on people as well as healthcare systems worldwide, influencing numerous elements of health and wellness plus health. Below we specify four substantial effects of diabetic issues mellitus:

6.3.1 Complication in Health

Diabetes mellitus significantly increases the risk of developing a range of health complications. Raised blood sugar levels over time can harm numerous body organs plus systems in the body, resulting in difficulties such as

Diabetic Person Retinopathy: Damage to the capillary in the retina results in vision disability as well as loss of sight.

Diabetic Person Nephropathy: Kidney damage leading to lowered kidney features as well as ultimate kidney failure.

Diabetic Person Neuropathy: Nerve damage creates discomfort, numbness, and prickling sensations, especially in the extremities.

Cardio Conditions: Diabetes poses a significant threat to cardiovascular disease, stroke, and peripheral artery disease, thereby increasing the risk of cardiovascular events and mortality.

These issues not only negatively impact lifestyle, but also increase the need for medical care and the associated costs.

6.3.2 Quality of Life Decreased

Maintaining diabetic issues typically requires significant lifestyle modifications, including dietary changes, regular monitoring of blood sugar levels, adherence to medication, and regular clinical visits. Managing diabetic issues can lead to tension, stress, and anxiety, which in turn can influence one's overall lifestyle. Issues such as vision loss, kidney failure, and nerve damage worsen the physical and psychological toll on people with diabetes mellitus and their households.

6.3.3 Economic Impact

The global healthcare expenditure on diabetes was estimated to be USD 760 billion in 2019-10% of total healthcare spending worldwide. Diabetes mellitus, of both type 1 and type 2, imposes a significant economic burden on individuals, healthcare systems, and society as a whole. There are direct healthcare costs related to medication, medical supplies, physician visits, hospitalizations, and management of complications. Indirect costs are related to productivity losses because of disability, premature mortality, and absenteeism from work.

6.3.4 Environmental Impact

The health authorities widely recognize diabetes mellitus as a high-demanding public health issue, and this trend of statistics is intensifying every year. Older and urban populations, sedentary lifestyles, unhealthy eating habits, and slowly penetrating obesity are the main reasons for the increased incidence of diabetes. When the diabetes flare-up is not addressed, the below raises complexity for the healthcare system and thus population wellbeing. Effective strategies to address this issue include health promotion, early disease prevention, immediate diabetes detection, and the implementation of high-yield management approaches.

6.4 LITERATURE REVIEW

Fico et al. [1] developed the personalized healthcare pathway (PHP) to manage diabetic disease, integrating it into an information and communication technology system to ease the shift from organization-centered to patient-centered care. We conducted small-scale exploratory research to test the platform.

Mougiakakou et al. [2] used the SMARTDIAB platform to manage diabetes. To manage type 1 diabetes, it integrates the most recent techniques in database (DB) technologies, communications, simulation algorithms, and data mining. Although it has equal software and hardware components for proper diabetes control, it lacks food recognition and an efficient meal recommendation.

Theis et al. [3] suggested a deep learning architecture to improve current methods of cruelty grading by incorporating the medical histories of diabetes patients. First, medical histories from previous hospital visits are transformed into event logs that can be used for process mining. After that, a process model that explains the patients' prior hospital experiences is found using the event logs. It is recommended to modify decline replay mining to forecast the in-hospital mortality of diabetic ICU patients by fusing known severity scores with demographic and medical data.

Sarathambekai et al. [4] have developed the best prediction model for machine learning-based diabetes guesses, based on an analysis of the prediction scores and outcomes of numerous research publications and articles. Based on the accuracy scores and other appraisal metrics acquired based on the prediction, common machine learning algorithms such as Decision Tree, random forest (RF) classifier, support vector machine (SVM), K-NN, Naïve Bayes, and the deep learning algorithm multi-layer perceptron classifier have remained evaluated from the studies.

El Bouhissi et al. [5] evaluated the most significant research on diabetes prediction and provided a method for predicting gestational diabetes, utilizing RF, SVM, and deep neural network (DNN) classifiers. Using a real dataset from Frankfurt Hospital, the experiment was carried out, showing that the RF method offers more accuracy.

Samir et al. [6] propose a system that combines deep learning-based human activity recognition (HAR) and food arrangement with non-invasive continuous glucose monitoring (CGM) strategies to track glucose levels. We establish a relationship between the pre-prandial and peak post-prandial glucose levels obtained from CGM and the glycemic load (GL) found in food, allowing us to estimate the rise in blood sugar levels and forecast hyperglycemia. Additionally, the technology links human activity to a drop in blood glucose levels, alerting users to potential hypoglycemia symptoms before they materialize.

Kim et al. [7] propose automated algorithms that combine deep learning and ensemble learning techniques to classify the images into four categories: diabetic macular edema (DME), normal, choroidal neovascularization (CNV), and drusen. For the classification, several convolutional neural networks (CNNs) are modified. Fully convolutional networks (FCN) are used to eliminate noise from training and test images, and a projection technique is employed to correct the images' tilted retinal layers. To further enhance performance, we train multiple CNNs and apply an ensemble learning method based on CNNs.

Santos et al. [8] propose a computer model to support medical diagnostics that uses pre-trained CNNs and can identify fundus grazes. Using the data-driven research diabetic retinopathy dataset, the model was refined, tested, and deployed using a YOLOv4 building and Darknet framework. It achieved a map of 11.13% and a mIoU of 13.98%.

Subha et al. [9] previously mentioned neural networks were used to develop deep learning models that are capable of measuring parameters including hemorrhages, blood vessels, microaneurysms, and fluid drop into several categories. The patient's eye's level of infection will be determined by the model. In particular, the kind and severity of the infection are determined using the fuzzy C-means algorithm. The degree of retinopathy can be determined using this paper. The entire system's user-friendly atmosphere makes identification easier. The interface will feature buttons that allow you to perform the required transformations on the provided test image.

Sazdov et al. [10] examined the use of federated learning (FL) in hospital readmission prediction for a 9-year dataset obtained from 130 US diabetes facilities. The findings imply that FL can retain data variety and privacy while achieving equivalent performance levels. This is a crucial component of FL since it permits ongoing, real-time learning without jeopardizing privacy.

6.5 PROPOSED METHODOLOGY

6.5.1 Diabetes Mellitus Diagnosis Early Detection Using Deep Learning

Early detection of diabetes mellitus would help reduce the complications, which later may cause suffering and thus lead to better patient outcomes. In recent years, the application of deep learning in medical imaging, electronic medical records, and genetic testing has shown improvement in outcomes using the analysis of various types of data. Deep learning algorithms, especially CNNs, which belong to such algorithms, have been recently used to make the previous detection of diabetes mellitus by investigating the images taken from medical sources, particularly retina imaging. Diabetic retinopathy is the number one complication of diabetes that is often responsible for the damage to the capillaries in the retina of the eye and may lead to sight-threatening vision decline when no proper treatment is provided. The use of deep learning models that can correctly detect diabetic retinopathy symptoms thanks to large-scale retinal image training is a great option for facilitating early diagnosis and prompt referral of treatment. With the help of these deep learning algorithm types, artificial intelligence can now diagnose retinopathy by analyzing retinal images and identifying specific lesions such as microaneurysms, hemorrhages, and exudates, which are typical of diabetic retinopathy. Deep learning models are more accurate due to the huge amount of image data available that they can detect; almost all of the abnormalities may not be noticed by regular human observations, thus deep learning models in healthcare will efficiently improve the diagnostic rate.

Moreover, deep learning methods recognize the analytical potential of multimodal data, e.g., demographic information, history of diseases, or lab results, to improve the robustness and accuracy of the early detection models. By combining data from various sources, deep learning technologies can recognize persons with a high chance of developing diabetes mellitus or its consequences. Consequently, this allows us to take preventive measures, thus minimizing the effects of these disorders (Figure 6.1).

The process for detecting diabetic mellitus early with deep learning techniques is shown in this image. Patient data is first gathered, and then the data is preprocessed to get it ready for model training. After training a deep learning model with the preprocessed data, the model's performance is assessed through evaluation. After the model has been trained and assessed, it can be used for new patient data to generate predictions, with

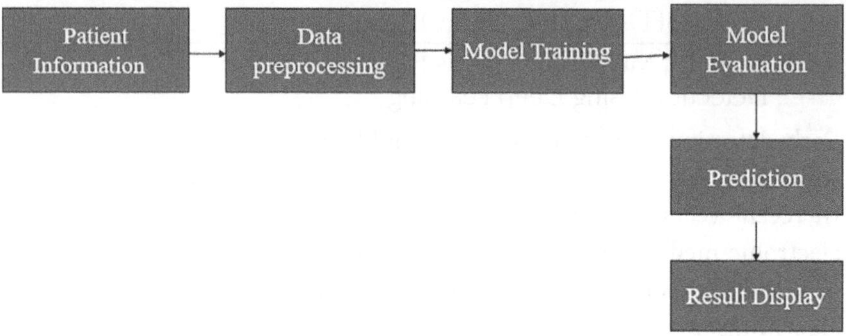

FIGURE 6.1 Early detection using deep learning.

the outcomes shown for decision-making and interpretation. Using deep learning, this method makes it possible to identify diabetes mellitus early on, which promotes proactive healthcare management and better patient outcomes.

6.5.2 Diabetes Treatment and Planning Using Deep Learning

Customized treatment and plans for diabetes mellitus focusing on patient precincts such as genetic aspects, daily living decisions, and medical records are interactive ways that have shown to be beneficial. Machine and deep learning methods will make improvements in future diabetes treatment since they are very good at analyzing large datasets and discovering trends that can then be used for personal care plan designs. Here's how deep learning can be utilized in diabetes treatment and planning: Here's how deep learning can be utilized in diabetes treatment and planning (Figure 6.2).

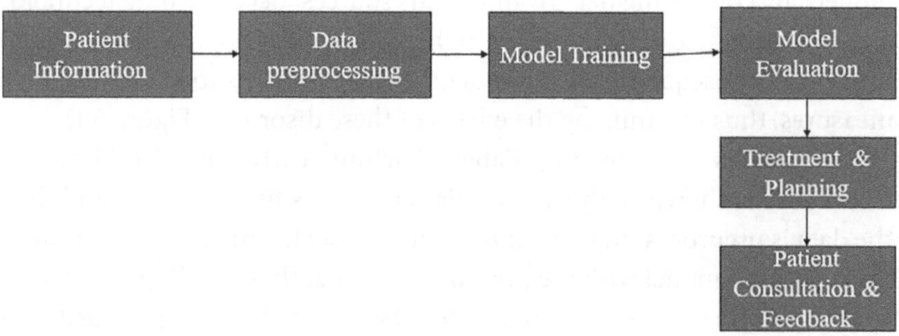

FIGURE 6.2 Treatment planning for diabetes mellitus in deep learning.

This figure presents a deep learning-based approach to treatment planning for diabetes mellitus. Healthcare practitioners can create individualized treatment regimens that maximize patient outcomes and enhance the quality of life by utilizing patient data and cutting-edge machine learning algorithms. Better diabetes mellitus management and increased patient satisfaction are the ultimate results of the iterative process of consultation and feedback, which makes sure that the treatment plan is continuously improved and modified to suit the patient's changing needs.

6.5.2.1 Personalized Treatment

The deep learning algorithms are capable of interpreting a variety of datasets on patients, encompassing genetic makeup, medical history, lifestyle, and glucose monitoring data into custom-made treatment plans. Integrating these data sources in deep learning models can allow disease status monitoring and identify the patterns and relationships that will help pick the precise drugs, dosages, and treatment schemes in individual patients. The cases, such as identifying a patient who is likely to have a better response to particular classes of anti-diabetic drugs or who may benefit from other dietary approaches or exercise schedules, can be made possible via deep learning.

6.5.2.2 Early Treatment

Deep learning approaches can detect early treatment responses and identify patients who may need revisions to their treatment programs. Deep learning algorithms can detect patterns of therapy success or failure by examining longitudinal patient data, such as changes in glucose levels, medication adherence, and clinical outcomes. This information can help healthcare practitioners intervene quickly to improve treatment efficacy and avoid problems.

6.5.2.3 Prediction

Deep learning algorithms can predict long-term results for diabetic patients based on their specific characteristics and treatment strategies. Deep learning models can uncover predictors of illness progression, complications, and mortality from vast DBs of patient outcomes, allowing healthcare providers to modify treatment approaches to reduce these risks. Deep learning techniques are useful tools for optimizing treatment and planning for diabetic patients by providing personalized care plans, optimizing glycemic control, assisting in treatment decision-making, facilitating early detection of treatment response, and predicting long-term outcomes. Deep learning has the potential to change diabetes care and improve patient outcomes by combining advanced computational methods with massive volumes of patient data.

6.5.3 Diabetes Complication Stages Using Deep Learning

Deep learning algorithms can assess a wide range of patient data, such as medical records, laboratory findings, and imaging investigations, to identify patients who are at high risk of developing diabetic problems. Deep learning algorithms can predict the possibility of problems such as diabetes retinopathy, nephropathy, neuropathy, and cardiovascular disease by finding tiny patterns and correlations in the data. Early diagnosis of high-risk individuals enables targeted therapies, such as intense glucose control, blood pressure management, and lifestyle changes, to avoid or delay the onset of problems. This technique can track the evolution of diabetic problems over time by assessing longitudinal patient data, such as changes in clinical parameters, biomarkers, and imaging findings. Deep learning models can notify healthcare professionals of the need for increased monitoring, therapy adjustments, or referral to specialty care by recognizing changes that indicate disease progression. Deep learning systems, for example, can examine retinal pictures to detect minor changes in the severity of diabetic retinopathy, allowing for early intervention to avert vision loss. Deep learning provides useful tools for predicting, monitoring, and controlling complications in various phases of diabetes mellitus. Deep learning has the potential to transform diabetes care by leveraging modern computational methods and massive amounts of patient data (Figure 6.3).

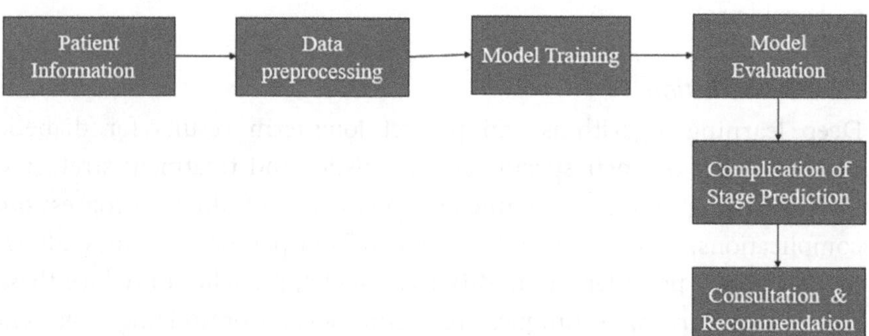

FIGURE 6.3 Diabetes mellitus complication stages using deep learning.

The complication stages of diabetes mellitus are predicted holistically using deep learning algorithms, as seen in this image. Through the utilization of classy machine learning algorithms and patient data, healthcare professionals can precisely forecast and handle diabetes mellitus problems, leading to improved patient outcomes and quality of life throughout. Patients' participation and observance of prescribed therapies are improved by the reiterative process of sessions and feedback, which guarantees that treatment programs are custom-made and patient-centered. Overall, the proactive organization of diabetes mellitus and its consequences is made easier by this all-inclusive strategy, which improves persistent outcomes in terms of health care.

6.6 RESULT AND ANALYSIS

When a patient clicks on the model exercise tab, the next shade appears. It displays the introduced data for training. The clicking model exercise creates and trains a deep learning approach (Figures 6.4 and 6.5).

The screen is used to gather and develop statistics for a patient recording form. The data is preprocessed on the software server based on previous assumptions. As an outcome of our attempts, we have a plan that can be tailored to the specific needs of individual patients (Figure 6.6).

To forecast diabetes, simply enter the appropriate patient information and click the site's "Predict Diabetes" option. When the "inform training set" tab is clicked, a new exercise set is created. After entering the patient's illness info and clicking the "predict diabetes" change, the patient's results

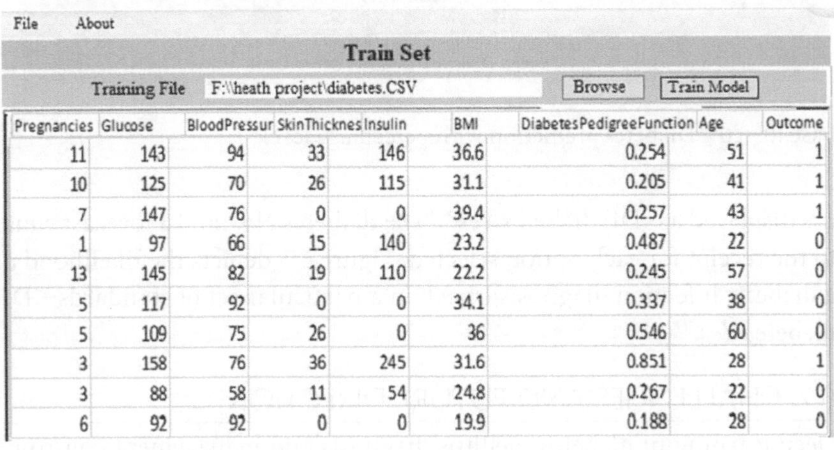

Pregnancies	Glucose	BloodPressur	SkinThicknes	Insulin	BMI	DiabetesPedigreeFunction	Age	Outcome
11	143	94	33	146	36.6	0.254	51	1
10	125	70	26	115	31.1	0.205	41	1
7	147	76	0	0	39.4	0.257	43	1
1	97	66	15	140	23.2	0.487	22	0
13	145	82	19	110	22.2	0.245	57	0
5	117	92	0	0	34.1	0.337	38	0
5	109	75	26	0	36	0.546	60	0
3	158	76	36	245	31.6	0.851	28	1
3	88	58	11	54	24.8	0.267	22	0
6	92	92	0	0	19.9	0.188	28	0

FIGURE 6.4 Training diabetes dataset.

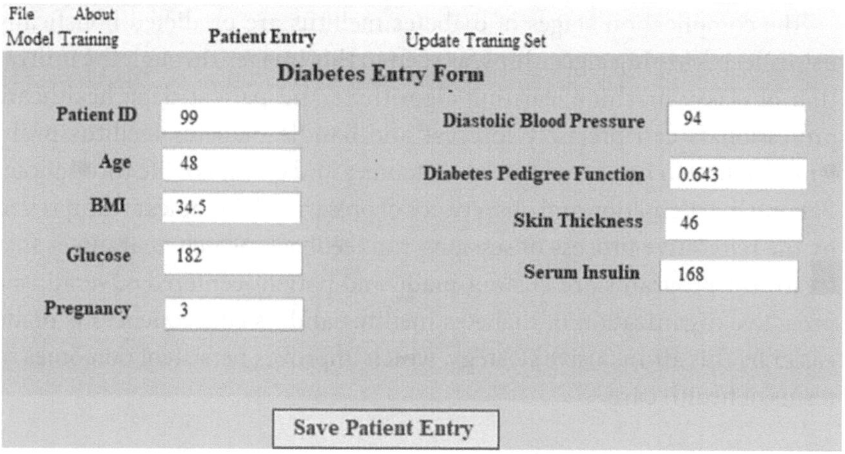

FIGURE 6.5 Diabetes prediction.

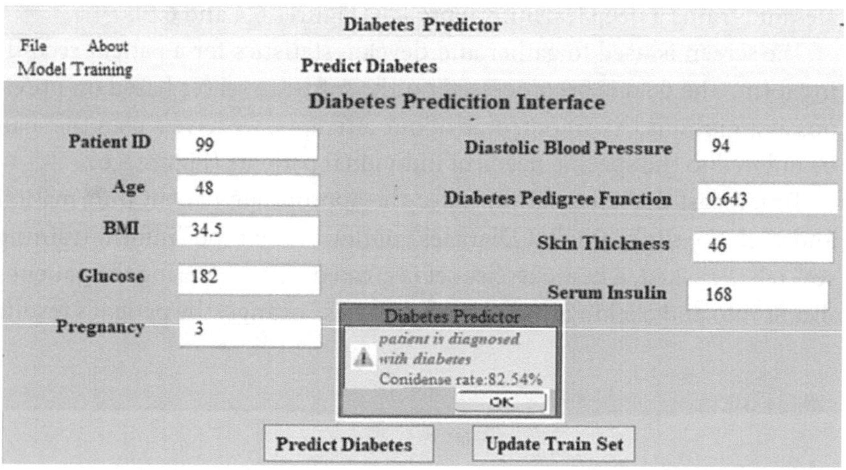

FIGURE 6.6 Diabetes prediction with percentage level.

are displayed as "D1: diabetes Yes," "D2: diabetes No," and a forecast equal to the receipt for each option selected. Figure 6.3 depicts the likelihood of a diabetic infection diagnosis based on a particular set of standards: "D1: diabetes Yes."

6.7 CHALLENGES AND FUTURE DIRECTION

Deep learning in diabetes mellitus diagnosis and management confronts several problems and uncertainties that must be addressed to reap full benefits. One key problem is the requirement for high-quality, standardized datasets when training deep learning models. Medical data, such as electronic health records and medical imaging, can contain inconsistencies, missing values, and heterogeneity between healthcare institutions, limiting the effectiveness and generalizability of deep learning algorithms. Furthermore, the interpretability and transparency of deep learning models are also an issue, especially in healthcare contexts where decisions can have a substantial impact on patient outcomes. Ensuring the explainability of deep learning predictions is critical for building trust with clinicians and patients as well as easing the use of these technologies in clinical practice. The future of deep learning in diabetes care looks promising. Continued research efforts to improve data quality, model interpretability, and regulatory frameworks will pave the path for the responsible and effective implementation of deep learning in clinical practice. Furthermore, cooperation among researchers, clinicians, policymakers, and industry stakeholders is critical for tackling these issues and realizing deep learning's transformative promise in diabetes diagnosis and management. Deep learning has the potential to change diabetes care, improve patient outcomes, and lower global healthcare costs if certain barriers are overcome.

6.8 CONCLUSION

Deep learning in diabetes mellitus diagnosis and management is a viable route for improving patient care while lowering healthcare expenditures. Deep learning techniques provide prospects to improve early identification, tailored treatment planning, and complication management in diabetes care, resulting in better results for patients with this chronic condition. Throughout this chapter, we've looked at how deep learning can help with diabetes care, such as early identification through medical imaging, personalized treatment planning based on patient data, and predictive modeling for complication management. Deep learning has shown the potential to transform diabetes diagnosis and management by leveraging advanced computational techniques and massive amounts of healthcare data. The use of deep learning in clinical practice is not without problems. Addressing concerns such as data quality, model interpretability, regulatory compliance, and ethical considerations will be critical to ensuring the responsible and effective use of deep learning technology in diabetes care. Through advanced computational algorithms and interdisciplinary teamwork, we can usher in a new era of personalized, data-driven healthcare that improves the lives of diabetes patients while simultaneously improving the efficiency and efficacy of healthcare delivery.

REFERENCES

1. G. Fico et al., "Integration of Personalized Healthcare Pathways in an ICT Platform for Diabetes Managements: A Small-Scale Exploratory Study," *IEEE Journal of Biomedical and Health Informatics*, vol. 20, no. 1, pp. 29–38, Jan. 2016, doi: 10.1109/JBHI.2014.2367863.
2. S. G. Mougiakakou et al., "SMARTDIAB: A Communication and Information Technology Approach for the Intelligent Monitoring, Management and Follow-up of Type 1 Diabetes Patients," *IEEE Transactions on Information Technology in Biomedicine*, vol. 14, no. 3, pp. 622–633, May 2010, doi: 10.1109/TITB.2009.2039711.
3. J. Theis, W. L. Galanter, A. D. Boyd and H. Darabi, "Improving the In-Hospital Mortality Prediction of Diabetes ICU Patients Using a Process Mining/Deep Learning Architecture," *IEEE Journal of Biomedical and Health Informatics*, vol. 26, no. 1, pp. 388–399, Jan. 2022, doi: 10.1109/JBHI.2021.3092969.
4. S. Sarathambekai et al., "A Comparative Analysis of Diabetes Prediction Models Using Machine Learning Algorithms," *2022 8th International Conference on Advanced Computing and Communication Systems (ICACCS), Coimbatore*, India, 2022, pp. 261–265, doi: 10.1109/ICACCS54159.2022.9785280.
5. H. El Bouhissi, R. E. Al-Qutaish, A. Ziane, K. Amroun, N. Yaya and M. Lachi, "Towards Diabetes Mellitus Prediction Based on Machine-Learning," *2023 International Conference on Smart Computing and Application (ICSCA)*, Hail, Saudi Arabia, 2023, pp. 1–6, doi: 10.1109/ICSCA57840.2023.10087782.
6. M. A. Samir, Z. A. Mohamed, M. A. A. Hussein and A. Atia, "Applying Deep Learning to Track Food Consumption and Human Activity for Non-intrusive Blood Glucose Monitoring," *2021 IEEE 12th Annual Ubiquitous Computing, Electronics & Mobile Communication Conference (UEMCON)*, New York, NY, 2021, pp. 0319–0324, doi: 10.1109/UEMCON53757.2021.9666662.
7. J. Kim and L. Tran, "Ensemble Learning Based on Convolutional Neural Networks for the Classification of Retinal Diseases from Optical Coherence Tomography Images," *2020 IEEE 33rd International Symposium on Computer-Based Medical Systems (CBMS)*, Rochester, MN, 2020, pp. 532–537, doi: 10.1109/CBMS49503.2020.00106.
8. C. Santos, M. S. de Aguiar, D. Welfer and B. M. Belloni, "Detection of Fundus Lesions through a Convolutional Neural Network in Patients with Diabetic Retinopathy," *2021 43rd Annual International Conference of the IEEE Engineering in Medicine & Biology Society (EMBC)*, Mexico, 2021, pp. 2692–2695, doi: 10.1109/EMBC46164.2021.9630075.

9. T. Subha, N. Susila, R. Ranjana, V. Priya and V. Nithyashree, "Analysis of Diabetic Retinopathy and Its Causes Using Fuzzy C-Means Model," *2021 4th International Conference on Computing and Communications Technologies (ICCCT)*, Chennai, India, 2021, pp. 527–532, doi: 10.1109/ICCCT53315.2021.9711840.
10. B. Sazdov et al., "Prediction of Hospital Readmission Using Federated Learning," *2023 30th International Conference on Systems, Signals and Image Processing (IWSSIP)*, Ohrid, North Macedonia, 2023, pp. 1–5, doi: 10.1109/IWSSIP58668.2023.10180282.

CHAPTER 7

Diabetes Mellitus Detection Using Deep Learning Model

M. Karthikeyan and R. Rengaraj

7.1 INTRODUCTION

The number of people with diabetes mellitus (DM) is high in developing countries like India owing to an increase in obesity and unhealthy lifestyles [1]. Type 2-DM can lead to microvascular and macrovascular problems. Recently, in India, these multiple organ problems have lead to an unconditional probability of death at an young age [1]. Type 2-DM is one of the most challenging threats to the Indian health care systems. The international diabetes federation estimated 74.9 million people with DM in 2021 with an expected projection of 124.9 million people with DM in 2045 [2]. According to the World Health Organization (WHO), people with age (≥45 and seldom exercising), body mass index (≥24 kg/m^2), impaired glucose tolerance or impaired fasting glucose, family history of DM, lower high-density lipoprotein cholesterol or hypertriglyceridemia, hypertension, or cardiovascular disease and gestation female (age ≥30) are at a high risk of DM [3]. The targets set by the WHO to achieve in the year 2030 are:

DOI: 10.1201/9781032708430-7

1. 80% of people with diabetes are to be diagnosed [4].
2. 80% of people with diagnosed diabetes should have good control of glycaemia [4].
3. 80% of people with diagnosed diabetes should have good control of blood pressure [4].
4. 60% of people with diabetes of 40 years or older must receive statins [4].
5. 100% of people with type 1 diabetes must have access to affordable insulin treatment and blood glucose self-monitoring [4].

The clinical detection of type 2-DM uses a fasting plasma glucose level greater than threshold. This glycaemic threshold levels may vary according to people belonging to a particular region because of the different glycaemic risk levels. Despite these difficulties, a decision support system for detecting type 2-DM is the need of the hour to reach a meaningful decision at a possible early stage. These constraints kindle the need for a simplified diagnostic process to detect type 2-DM.

With the latest technological development and the impact of artificial intelligence (AI) in every day activities, development of machine learning (ML) algorithms helps in making a quick decision support system in type 2-DM detection. Several neural network-based algorithms [5], deep neural networks [6,7], and conventional ML models [8–17] are used in the research of type 2-DM detection.

Considering the challenges and accuracy rate in type 2-DM detection available in the literature and the usage of modern AI tools, there is vital need for a new improved model. Hence, in this chapter, we propose a robust deep learning model (DLM) for type 2-DM detection to create an efficient decision support system. The chapter's contributions can be outlined as follows:

1. We propose a simple DLM for type 2-DM detection. We also explore the ML algorithm with Bayesian optimization for hyperparameter tuning. A comparison is made in terms of performance metrics for the algorithms considered.
2. The model undergoes training using a dataset from Germany hospital dataset and is subsequently tested with PIMA Indian diabetes

dataset (PIDD) showing the generalization supremacy of the proposed model with the available ML.

3. The concept of transfer learning is strengthened by training the model with one dataset and testing it with another dataset that was not part of the training process.

The remainder of the chapter is structured as follows: Section 7.2 presents the literature on DM prediction. Section 7.3 introduces the methodology. Section 7.4 presents the proposed DLM. Section 7.5 presents the experimental dataset. Results and discussions are explained in Section 7.6. Section 7.7 concludes the chapter.

7.2 RELATED WORK

In recent years, the field of AI and ML has played a vital role to detect the likelihood of type-2 DM disease. Many algorithms and tools have been developed showing the incredible potential of this research domain. We limit our review to recent findings which are closely related to the issue presented. Neural network-based prediction shows improvement in prediction accuracy recently in DM research. In Ref. [6], authors utilized min-max normalization and autoencoder to resolve data normalization, imbalance in dataset, and feature augmentation. Multilayer perceptron was used for classification to achieve a 92.31% accuracy [6]. In Ref. [5], authors designed a classifier based on artificial backpropagation scaled conjugate gradient neural network with 93% accuracy without data preprocessing. An improved neural network-based method was reported in Ref. [18] using *k*-nearest neighbour to achieve an F1-score of 98%. Authors of Ref. [19] applied Pearson correlation and median value imputation for feature selection. Their deep neural network model achieved 88.6% accuracy [18].

The ML classification model yielded good prediction accuracies when compared to neural network-based methods. In Ref. [20], authors evaluated different ML classifiers: light gradient boosting machine (LGBM), linear regression (LR), and support vector machines (SVMs) with LGBM outperforming other models with an accuracy of 86%. In Ref. [18], classification accuracies of ML algorithms decision tree (DT), random forest (RF), naïve bayes (NB), *k*-nearest neighbour (KNN), and adaptive boosting (AB) were compared. Authors of Ref. [7] developed a deep neural network model with an accuracy of 98.07%. In Ref. [21], feature selection methods such as forward and backward feature selection strategy were employed to improve the classification accuracies. In Ref. [22], Pearson correlation and mean value imputation were used for feature selection and missing value imputation with grid search for hyper parameter tuning. Different classification models: extreme boosting (XB), AB, RF, and K-NN were experimented in Ref. [21] and achieved 94.6% accuracy for XB.

Authors of Ref. [23] proposed a twice-growth deep neural network (2GDNN) and achieved 97% accuracy for PIMA and laboratory of the medical city hospital dataset. Authors of Ref. [8] used two datasets, PIMA Indian dataset and Germany hospital dataset, to design an ML-based classifier. For this, different ML algorithms were chosen, i.e., SVM, NB, KNN, RF, LR, and DT [8]. The findings revealed that for PIMA dataset, SVM outperformed with an accuracy of 74%, whereas KNN and RF performed better with an accuracy of 98.7% for Germany dataset [8]. The above related works are summarized in Table 7.1. A detailed review about related work in ML and deep learning-based classification model and a systematic review about DM is available in Refs. [9–17,24].

TABLE 7.1 Summary of Previous Milestone Classification Model

Authors	**Classification Model Type**	**Remarks**
María Teresa García-Ordás et al. [6]	Multilayer perceptron	Missing values were removed and achieved 92.318%
Muhammad Mazhar Bukhari et al. [5]	Artificial neural network scaled conjugate gradient neural network	Attained 93% accuracy
Kumarmangal Roy et al. [18]	KNN	Attained 98% accuracy
Jobeda Jamal Khanam et al. [19]	Multilayer perceptron and deep learning with two hidden layers	Attained 98% accuracy
Talha Mahboob Alam et al. [20]	LGBM	Achieved 86% accuracy
Naz and Ahuja [7]	Deep learning approach	Achieved 98.07% accuracy
Sivaranjani et al. [21]	RF, SVM	Feature selection and dimensionality reduction with 83% accuracy
Hasan et al. [22]	XB, AB, RF, DT, and KNN	Achieved 94.6% accuracy
Kirti Kangra [8]	SVM, NB, KNN, RF, LR, and DT	Achieved 98.7% accuracy
Zaigham Mushtaq [9]	Voting-based classifier	Achieved 82% accuracy

7.3 METHODOLOGY

Figure 7.1 shows the diagrammatic illustration of the methodology carried out for DM detection. The methodology consists of data preprocessing for missing values detection [23]. The dataset is divided into three categories: Training, validation, and testing. *Half of the data is allocated for training, 30% is reserved for tenfold cross validation, and remaining 20% is utilized to test the developed model as shown in Figure 7.2.* Finetuning the model is carried out by hyperparameter tuning for ML algorithms. A classification model based on ML algorithms such as DT, LR, NB, SVM, KNN, and ensemble learning algorithm (ELA) is used for DM detection in this work. The hyperparameter of these ML classification models is tuned using Bayesian optimization. The key advantage of this optimization is that in each iteration the parameter selection is based on previous iteration. Instead of choosing the parameters randomly in each stage, this optimization chooses from the relevant search space and discards the search space

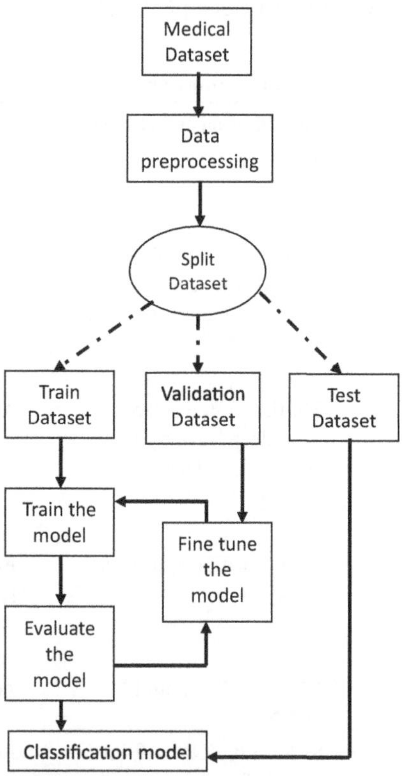

FIGURE 7.1 Diagrammatic illustration of the methodology.

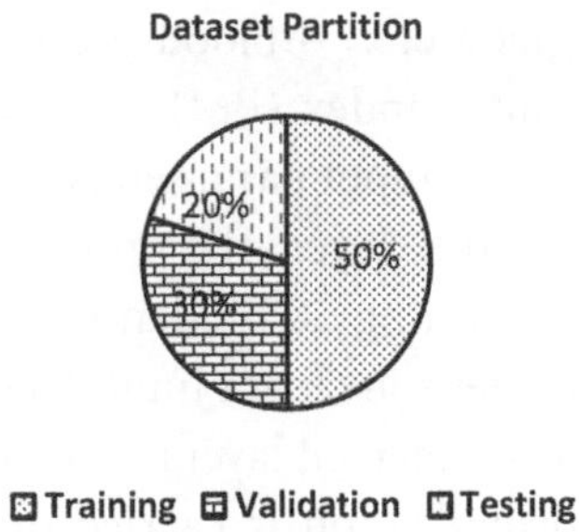

FIGURE 7.2 Dataset partition strategy.

that does not deliver a promising solution. Hence, it requires less iteration than other tuning methods like random search or grid search. This increases the validation accuracy and reduces the tuning time as compared to the same model optimized by random or grid search approach for hyperparameter tuning [25].

7.4 DEEP LEARNING MODEL (DLM)

The architecture of the proposed DLM for DM prediction is shown in Figure 7.3. It consists of input feature layer followed by fully connected layers with batch normalization and rectified linear unit (ReLU) activation. Finally, softmax layer and classification layer are available in the proposed DLM. The dataset contains eight input features represented

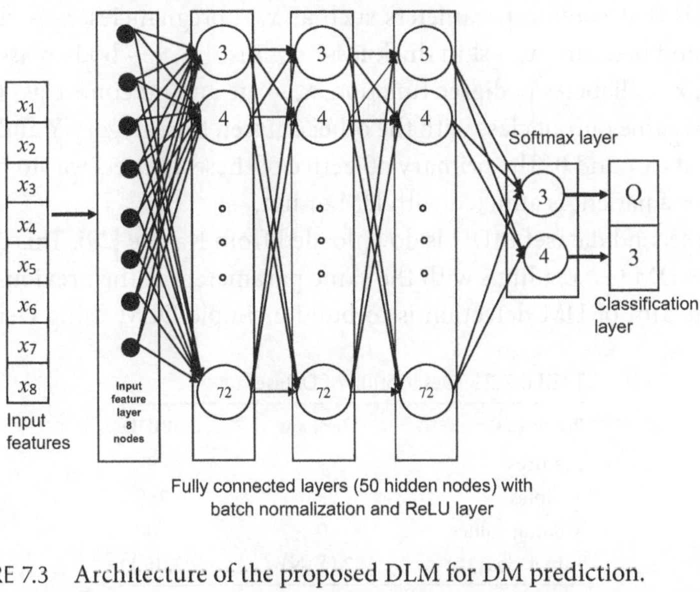

FIGURE 7.3 Architecture of the proposed DLM for DM prediction.

x_1 – pregnancies, x_2 – glucose, x_3 – blood pressure, x_4 – skin thickness, x_5 – insulin, x_6 – body mass index (BMI), x_7 – diabetes pedigree function, and x_8 – age. These eight input parameters of each medical record are given as input for input feature layer. The proposed model has three fully connected layers each containing 50 hidden neurons. From Figure 7.3, it is understood that all the neurons from input feature layer are connected to all neurons of the fully connected layers. This scheme continues till the softmax layer. Batch normalization is carried out to make the network faster and stable. Batch normalization works by normalizing the output of a previous activation layer by subtracting the batch mean and dividing by the batch standard deviation [26]. After this step, the result is then scaled and shifted by two learnable parameters, gamma and beta, which are unique to each layer [26]. A ReLU is an activation function that introduces the property of nonlinearity to a DLM and solves the vanishing gradients issue of the neural network [27]. The output of the fully connected layer is given to the softmax layer that gives the probability score of each class. Based on this probability, the output of the classification layer is category '1' or '0'. Category '1' represents the DM and '0' represents non-DM.

7.5 DATASET

Two similar datasets as shown in Table 7.2 are downloaded to develop and test the DLM. The dataset shared by the hospital from Frankfurt, Germany, is downloaded from Kaggle [28]. The dataset has 2000 DM test readings that contain parameters such as x_1 – pregnancies, x_2 – glucose, x_3 – blood pressure, x_4 – skin thickness, x_5 – insulin, x_6 – body mass index (BMI), x_7 – diabetes pedigree function, x_8 – age and outcome class (Y, N). To have same output class with the other dataset, the category Y and N are changed to 1 and 0. The primary objective of these datasets was to predict whether a patient is afflicted with DM or not.

The second dataset PIDD is downloaded from Kaggle [29]. This dataset has 768 DM test readings with the same parameters as the previous dataset. The aim of DM detection is to build a simple DLM using Germany

TABLE 7.2 Description of Dataset

Properties	Germany	PIDD
Features	9	9
Samples	2000	768
Missing values	0	0
Target class	2 (Y, N)	2 (0, 1)

dataset for training and validation. The developed deep learning classification model is tested with PIDD that are unseen by the model during training and validation. It is a kind of transfer learning where the trained DLM is reused for a new dataset.

The performance metrics utilized for comparing the various classification models are classification accuracy (CA), precision (P), recall (R), and F1-score (F_1). Equations (7.1)–(7.4) give the performance metrics of the classification models. Here, true positive (TP) represents the number of persons with DM. True negative (TN) represents the number of persons without DM. False positive (FP) represents the number of persons where the predicted output class is DM, whereas the actual output class is no DM. False negative (FN) represents number of records where the predicted output class is no DM, whereas the actual output class is DM.

$$\mathrm{CA} = \frac{\mathrm{TP}+\mathrm{TN}}{\mathrm{TP}+\mathrm{FP}+\mathrm{TN}+\mathrm{FN}} * 100 \tag{7.1}$$

$$P = \frac{\mathrm{TP}}{\mathrm{TP}+\mathrm{FP}} \tag{7.2}$$

$$R = \frac{\mathrm{TP}}{\mathrm{TP}+\mathrm{FN}} \tag{7.3}$$

$$F_1\text{-score} = 2 * \frac{\text{precision} * \text{recall}}{\text{precision} + \text{recall}} \tag{7.4}$$

7.6 RESULTS AND DISCUSSION

7.6.1 Training

Classification models are trained using Germany hospital dataset and its performance is assessed using tenfold cross validation. Figure 7.4 illustrates the training and validation accuracy of the proposed DLM. Figure 7.5 depicts the training and validation loss of the proposed DLM. To show the robustness of the proposed DLM, a comparison is made with ML algorithms. The hyperparameters of these ML algorithms are tuned by Bayesian optimization. A minimum of 30 objective functions evaluation were completed for all ML algorithms during hyperparameter tuning. Table 7.3 presents the accuracy comparison of the proposed DLM with the

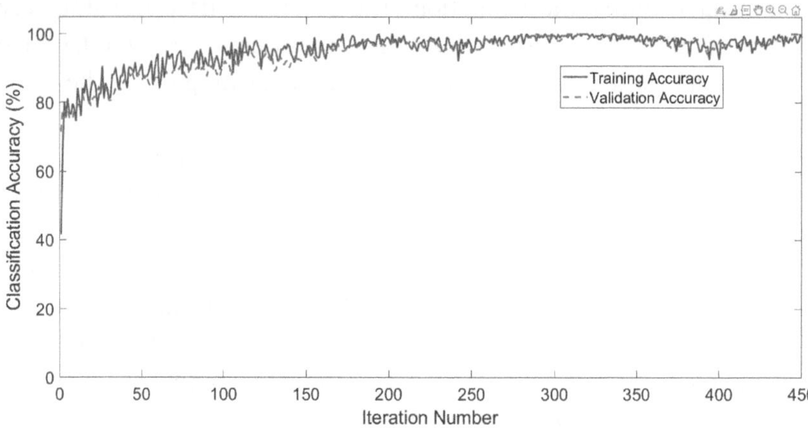

FIGURE 7.4 Training and validation accuracy of the proposed DLM.

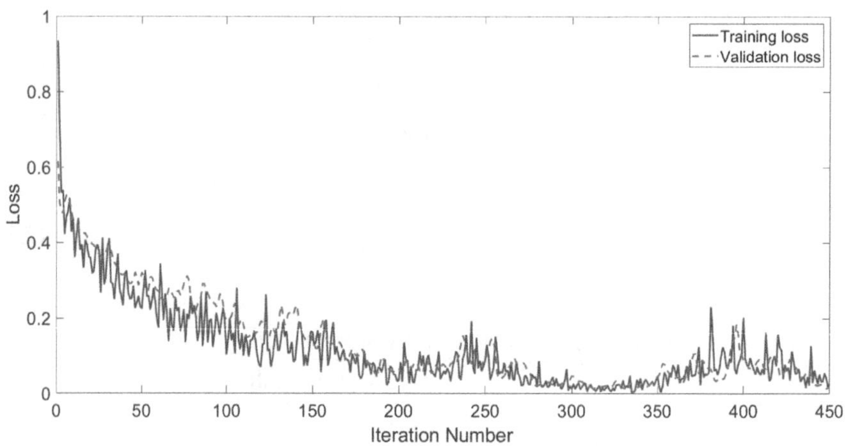

FIGURE 7.5 Training and validation loss of the proposed DLM.

TABLE 7.3 Validation Accuracy for the Germany Hospital Dataset

Classification Model	TP	TN	FP	FN	Validation Accuracy (%)
Decision tree	609	1233	83	75	92.1
Logistic regression	382	1170	146	302	77.6
Naïve bayes	339	1178	138	345	75.8
SVM	645	1316	0	39	98
KNN	661	1303	13	23	98.2
Ensemble learning	676	1280	36	8	97.8
Proposed DLM	684	1316	0	0	100

other ML algorithms. Also, TP, TN, FP, and FN values are found to be better compared to ML algorithms. Figure 7.6 shows the receiver operating characteristics (ROC) curve of ML algorithms considered here. The area under the curve (AUC) is good for DT, SVM, KNN, and ELA. Table 7.4 displays the comparison of performance metrics among the various classification models during training. The proposed DLM outperforms other ML models and results in literature [8] for the same dataset.

Table 7.4 displays the comparison of performance metrics among the various classification models during training.

7.6.2 Testing

The classification models trained using Germany hospital dataset is stored. These models are tested with PIDD. PIDD dataset is a completely new medical record for the model as the dataset is not utilized during training. This is done to show the robustness, effectiveness of developed model towards reliable detection accuracy, and generalization and adaptability of the proposed model. The test accuracy results and performance metrics comparison are shown in Tables 7.5 and 7.6. The proposed DLM outperforms other ML models and results in literature [8,23] for the same dataset. The proposed DLM failed to classify ten samples. These ten sample numbers and their anomalous parameter are shown in Table 7.7. The parameters listed in Table 7.7 have irregular values which may be an obstacle from achieving 100% testing accuracy. The classification testing accuracy for the PIDD dataset was 98.7% which proves the robustness, reliability, and effectiveness of the proposed DLM.

7.6.3 Future Work

The scope for future work in DM detection using deep learning has several areas for further exploration. The key areas include improved accuracy for a large dataset, early prediction and prevention, integrating with edge computing-based wearable devices, and deployment with a cheapest price affordable by all. To explore the DLM further like implementation with pretrained model, there is a need for a large dataset. Once a large dataset collected from across the various geographical location is available, using transfer learning-based DLMs are possible. Also, there is further scope to study the detection of DM in children and teenagers. Then, the performance of the proposed DLM can be investigated further. The goal will be to develop a simple model for detecting DM irrespective of gender, age, or race.

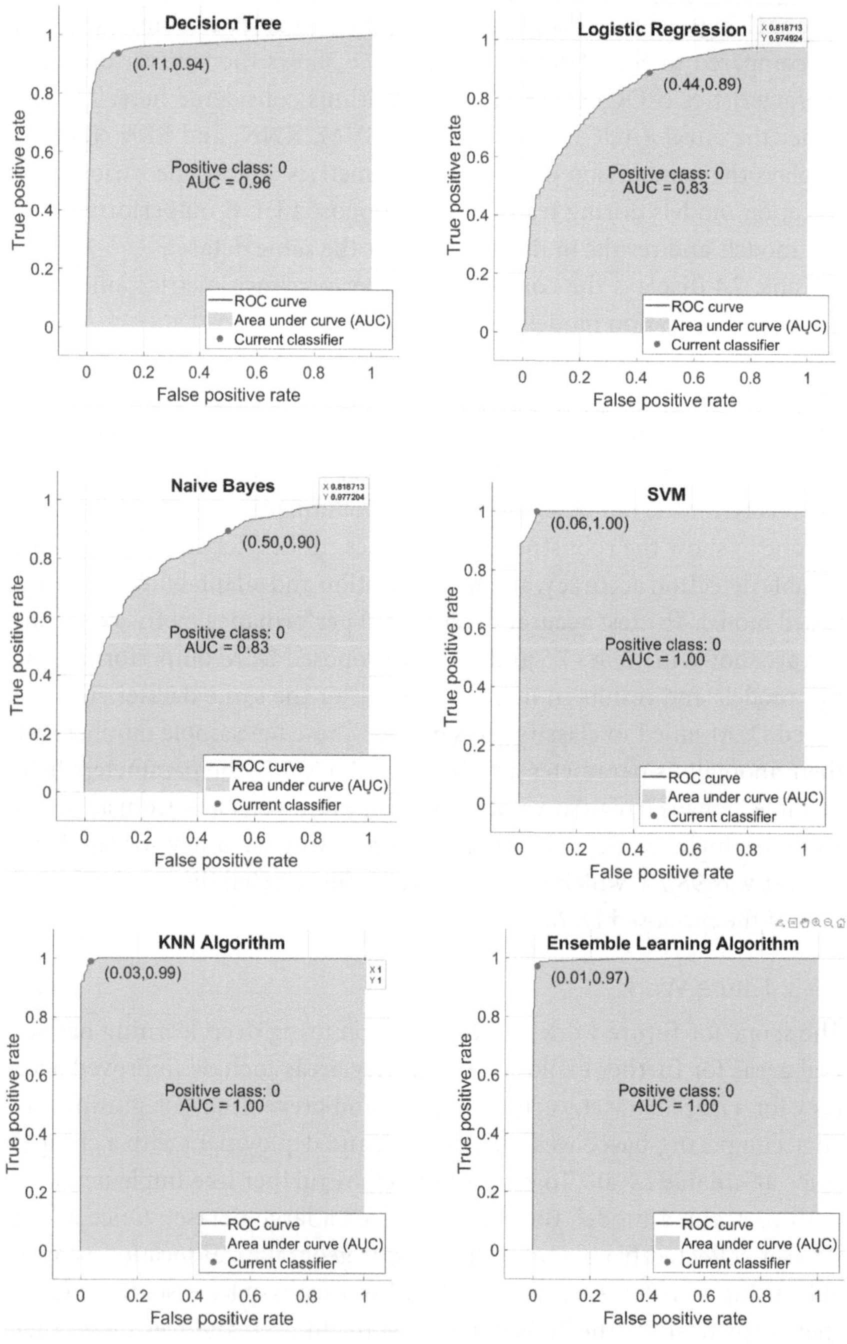

FIGURE 7.6 (a) –(f) ROC curves of different ML algorithms.

TABLE 7.4 Performance Metrics Comparison of Different Classification Models on Germany Hospital Dataset

Classification Model	Precision (%)	Recall (%)	F1-Score (%)
Decision tree	88.01	89.04	88.52
Logistic regression	72.35	55.85	63.04
Naïve bayes	71.07	49.56	58.40
SVM	100.00	94.30	97.07
KNN	98.07	96.64	97.35
Ensemble learning	94.94	98.83	96.85
2GDNN [23]	97.35	96.67	96.97
Proposed DLM	100.00	100.00	100.00

TABLE 7.5 Testing Accuracy for the PIDD

Classification Model	TP	TN	FP	FN	Testing Accuracy (%)
Decision tree	245	492	8	23	96
Logistic regression	152	449	51	116	78.3
Naïve bayes	124	453	47	144	75.1
SVM	249	500	—	19	97.5
KNN	252	494	8	14	97.14
Ensemble learning	256	496	6	10	97.92
Proposed DLM	261	497	3	7	98.7

TABLE 7.6 Performance Metrics Comparison of Different Classification Models on PIDD

Classification Model	Precision (%)	Recall (%)	F1-Score (%)
Decision tee	96.84	91.42	94.05
Logistic regression	74.88	56.72	64.54
Naïve bayes	72.51	46.27	56.49
SVM	100.00	92.91	96.32
KNN	96.92	94.74	95.82
Ensemble learning	97.71	96.24	96.97
2GDNN [23]	97.34	97.25	99.01
Proposed DLM	98.86	97.39	98.12

7.7 CONCLUSION

In this chapter, we proposed a simple DLM to detect DM. The DLM has exhibited remarkable accuracy in detecting DM for the Germany hospital dataset and PIDD. The hyperparameter tuning of the ML algorithms was conducted using Bayesian optimization. The proposed model was trained

TABLE 7.7 Details of Misclassified Sample Numbers

Sl. No.	Sample Number	Actual Output Class	Predicted Output Class	Abnormal Parameter
1.	8	0	1	BMI
2.	21	0	1	BMI
3.	31	0	1	BMI
4.	7	1	0	Glucose, blood pressure, BMI
5.	10	1	0	Glucose, diabetic pedigree function
6.	14	1	0	BMI, diabetic pedigree function
7.	16	1	0	Glucose, BMI
8.	25	1	0	Diabetic pedigree function
9.	221	1	0	Blood pressure
10.	398	1	0	Blood pressure

using Germany hospital dataset with tenfold cross validation and tested with PIDD. Training and test results proved the strength of the proposed model compared with ML algorithms. With the growing deployment of deep learning algorithms in real time along with healthcare data, there exists a bright future for DM detection and diagnosis.

REFERENCES

1. R. Pradeepa, V. Mohan, Epidemiology of type 2 diabetes in India, *Indian J Ophthalmol.* 69(11), 2932–2938 (2021). https://doi.org/10.4103/ijo.IJO_1627_21.
2. I. D. Federation, *IDF Diabetes Atlas,* 10th Edition. Preprint at (2021). International Diabetes Federation, Brussels, Belgium
3. H. Wu, S. Yang, Z. Huang, J. He, X. Wang, Type 2 diabetes mellitus prediction model based on data mining, *Inform Med Unlocked*, 10, 100–107 (2018). ISSN 2352–9148, https://doi.org/10.1016/j.imu.2017.12.006.
4. Seventy-Fifth World Health Assembly, *Geneva.* 22–28 May 2022: *Resolutions and Decisions, Annexes.* Geneva: World Health Organization (2022) (WHA75/2022/REC/1; https://iris.who.int/handle/10665/365610).
5. M. M. Bukhari, B. F. Alkhamees, S. Hussain, A. Gumaei, A. Assiri, S. S. Ullah, An improved artificial neural network model for effective diabetes prediction, *Complexity*, 2021, Article ID 5525271, 10 pages (2021). https://doi.org/10.1155/2021/5525271.
6. M. T. García-Ordás, C. Benavides, J. A. Benítez-Andrades, H. Alaiz-Moretón, I. García-Rodríguez, Diabetes detection using deep learning techniques with oversampling and feature augmentation, *Comp Methods Programs Biomed*, 202, 105968 (2021). https://doi.org/10.1016/j.cmpb.2021.105968.
7. H. Naz, S. Ahuja, Deep learning approach for diabetes prediction using PIMA Indian dataset. *J Diabetes Metab Disord*, 19, 391–403 (2020). https://doi.org/10.1007/s40200-020-00520-5.

8. K. Kangra, J. Singh, Comparative analysis of predictive machine learning algorithms for diabetes mellitus, *Bull Electr Eng Inform*, 12(3), 1728–1737 (2023). ISSN:2302–9285, https://doi.org/10.11591/eei.v12i3.4412.
9. Z. Mushtaq, M. F. Ramzan, S. Ali, S. Baseer, A. Samad, M. Husnain, Voting classification-based diabetes mellitus prediction using hypertuned machine-learning techniques, *Mobile Inform Sys*, 2022, 16 (2022). https://doi.org/10.1155/2022/6521532.
10. V. Rawat, S. Joshi, S. Gupta, D. P. Singh, N. Singh, Machine learning algorithms for early diagnosis of diabetes mellitus: A comparative study, *Mater Today Proc*, 56(part 1), 502–506 (2022). https://doi.org/10.1016/j.matpr.2022.02.172.
11. L. Ismail, H. Materwala, M. Tayefi, P. Ngo, A. P. Karduck, Type 2 diabetes with artificial intelligence machine learning: Methods and evaluation, *Arch Comput Methods Eng*, 29(1), 313–333 (2022). https://doi.org/10.1007/s11831-021-09582-x.
12. R. Patra, B. Khuntia, Analysis and prediction of Pima Indian diabetes dataset using SDKNN classifier technique, *IOP Conf Ser Mater Sci Eng*, 1070(1), 1–14 (2021). https://doi.org/10.1088/1757–899X/1070/1/012059.
13. K. S. Kumari, K. Bhargavi, Performance analysis of diabetes mellitus using machine learning techniques, *Turkish J Comput Math Educ*, 12(6), 225–230 (2021).
14. F. Alaa, A. M. Al-bakry, Diagnosis of diabetes using machine learning algorithms, *Mater Today Proc*, (2021). https://doi.org/10.1016/j.matpr.2021.07.196.
15. S. Rajendar, R. Thangaraj, J. Palanisamy, V. K. Kaliappan, Comparative analysis of classifier models for the early prediction of type 2 diabetes, *Int J Adv Sci Technol*, 29(7), 2184–2194 (2020).
16. R. Patil, S. Tamane, S. A. Rawandale, K. Patil, A modified mayfly-SVM approach for early detection of type 2 diabetes mellitus, *Int J Electr Comput Eng*, 12(1), 524–533 (2022), https://doi.org/10.11591/ijece.v12i1.pp524–533.
17. S. Samreen, Memory-efficient, accurate and early diagnosis of diabetes through a machine learning pipeline employing crow search-based feature engineering and a stacking ensemble, *IEEE Access*, 9, 134335–134354 (2021). https://doi.org/10.1109/ACCESS.2021.3116383.
18. K. Roy, M. Ahmad, K. Waqar, K. Priyaah, J. Nebhen, S. S. Alshamrani, M. A. Raza, I. Ali, An enhanced machine learning framework for type 2 diabetes classification using imbalanced data with missing values, *Complexity*, 2021, 21 (2021). https://doi.org/10.1155/2021/9953314.
19. J. J. Khanam, S. Y. Foo, A comparison of machine learning algorithms for diabetes prediction, *ICT Express*, 7(4), 432–439 (2021). https://doi.org/10.1016/j.icte.2021.02.004.
20. T. M. Alam, M. A. Iqbal, Y. Ali, A. Wahab, S. Ijaz, T. I. Baig, A. Hussain, M. A. Malik, M. M. Raza, S. Ibrar, Z. Abbas, A model for early prediction of diabetes, *Inform Med Unlocked*, 16, 100204 (2019). https://doi.org/10.1016/j.imu.2019.100204.

21. S. Sivaranjani, S. Ananya, J. Aravinth, R. Karthika, Diabetes prediction using machine learning algorithms with feature selection and dimensionality reduction, 2021 7th International Conference on Advanced Computing and Communication Systems (ICACCS), Coimbatore, India, 2021, pp. 141–146, https://doi.org/10.1109/ICACCS51430.2021.9441935.
22. M. K. Hasan, M. A. Alam, D. Das, E. Hossain, M. Hasan, Diabetes prediction using ensembling of different machine learning classifiers, *IEEE Access*, 8, 76516–76531 (2020). https://doi.org/10.1109/ACCESS.2020.2989857.
23. C. C. Olisah, L. Smith, M. Smith, Diabetes mellitus prediction and diagnosis from a data preprocessing and machine learning perspective, *Comp Methods Programs Biomed*, 220, 106773 (2022). https://doi.org/10.1016/j.cmpb.2022.106773.
24. J. Chaki, S. Thillai Ganesh, S. K. Cidham, S. Ananda Theertan, Machine learning and artificial intelligence based Diabetes Mellitus detection and self-management: A systematic review, *J King Saud Univ Comput Inform Sci*, 34(6, Part B), 3204–3225 (2022). https://doi.org/10.1016/j.jksuci.2020.06.013.
25. A. H. Victoria, G. Maragatham, Automatic tuning of hyperparameters using Bayesian optimization. *Evol Syst*, 12, 217–223 (2021). https://doi.org/10.1007/s12530-020-09345-2.
26. S. Ioffe, C. Szegedy, Batch normalization: Accelerating deep network training by reducing internal covariate shift. arXiv:1502.03167 (2015).
27. K. Hara, D. Saito, H. Shouno, Analysis of function of rectified linear unit used in deep learning, 2015 International Joint Conference on Neural Networks (IJCNN), Killarney, Ireland (2015), pp. 1–8. https://doi.org/10.1109/IJCNN.2015.7280578.
28. Kaggle, *Diabetes*. https://www.kaggle.com/datasets/johndasilva/diabetes (accessed Feb. 24, 2024).
29. Kaggle, *Pima Indians Diabetes Database*. https://www.kaggle.com/uciml/pima-indians-diabetes-database (accessed Feb. 24, 2024).

CHAPTER 8

A Comprehensive Review of the Use of Deep Learning Algorithms in Diabetes Mellitus Detection and Diagnosis

Katarzyna Wiltos

8.1 INTRODUCTION

Diabetes is a persistent metabolic disease that hinders the body's ability to produce insulin in the pancreas causing abnormal glucose levels in the blood (hyperglycemia). As blood glucose levels remain high, the blood vessels become affected and eventually damaged. Insulin is crucial for the proper metabolism of fat and protein [1]. If left untreated, the insufficient level of insulin can consequently lead to serious health complications such as heart disease, obesity, stroke, kidney failure, foot-related complications, vision impairment, nerve damage, or cognitive problems [2]. The complexity of the disease is noticeable through the amount of different types of diabetes with the most common ones being type 1, type 2, and gestational diabetes [3]. Type 1 diabetes is an autoimmune disease, caused

DOI: 10.1201/9781032708430-8

by one's immune system destroying cells responsible for producing insulin (pancreatic β-cells) [4]. Type 2 diabetes, which constitutes over 90% of all diabetes worldwide [1], is not an autoimmune disease, but a condition where the body produces insulin while not using it properly or its levels are insufficient [5]. Gestational diabetes develops during the second or third trimester of pregnancy and usually resolves after delivery [6].

Nonetheless, the occurrence of gestational diabetes increases the risk of developing type 2 diabetes later in life [1].

As the number of affected people is high and is continually rising worldwide, the research area of detecting diabetes mellitus at its early stages holds great importance [7]. Thus, it is essential to provide a comprehensive review of optimal techniques and identify the latest top-performing approaches. The following sections will concern the most recent deep learning approaches in diabetes mellitus detection and their possible practical applications in the healthcare system. It is estimated that around 537 million adults aged 20–70 years are living with the disease. This number is expected to rise further to 643 million (11.3%) by 2030 and to 783 million (12.2%) by 2045. Additionally, around 240 million people are believed to be living undiagnosed without any medical treatment applied [1] (Figure 8.1).

Development and ongoing enhancement of possible treatment options, along with early detection of diabetes mellitus, are crucial for minimizing the impact of the disease on the health and quality of life of chronically ill

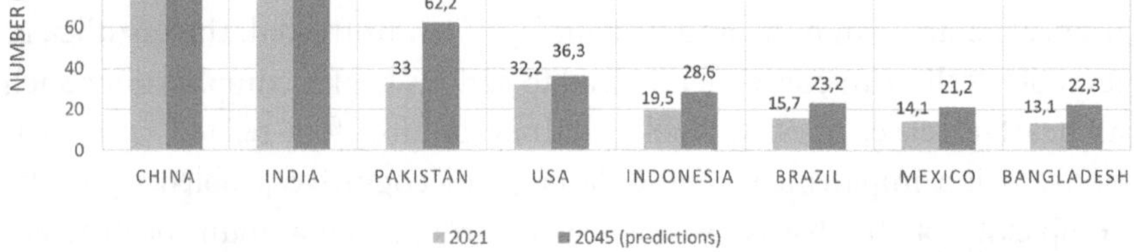

FIGURE 8.1 A graph presenting countries with the highest prevalence of diabetes mellitus in millions for 2021 and predictions for 2045. (The data used to create this *figure were obtained from the table in Ref. [1]. The data were processed and visualized by the author.)

patients [8]. Early detection is vital for ensuring as little health destruction as possible and preventing the disease from causing further harm to the body, ultimately contributing to patients' health-related life satisfaction [9]. Improvement of diagnosis methods is critical as it is estimated that up to one-half of type 2 diabetes remains undiagnosed causing many irreversible long-term health impairments, severe disabilities, and even death [1].

The diagnosis procedure is usually based on testing blood glucose levels. According to the World Health Organization (WHO), a fasting blood glucose level equal to or exceeding 126 mg/dL (hyperglycemia) indicates diabetes [10]. However, deep learning algorithms simultaneously consider more insightful data such as medical records, weight, age, heart rate variability, genetics, retinal images, physique index, pulse rate [11], or electrocardiogram information [12]. This approach allows succinctly precise diagnosis and may serve as an independent decision tool or assist greatly in human review. These algorithms can analyze large amounts of complex data and can identify more subtle relations between data features providing more in-depth results [13].

8.2 RECENT ADVANCES IN DIAGNOSTIC METHODS

Artificial intelligence techniques not only allow for effective early detection of new-onset cases but also serve as a robust process of therapeutic monitoring of the disease in real time [6]. Among these methods, deep learning techniques have the potential to significantly advance the field of diabetes detection, prediction, and management by providing more robust and efficient methods for identifying patients at risk and enabling the development of more personalized treatment plans. Implementing these methods in practice is beneficial for the diagnosis process, as massive and complex datasets can be analyzed more effectively than with traditional machine learning approaches. Deep learning methods can provide more precise results as they are capable of identifying patterns not evident to human review. Another advantage is the ability to work with unstructured and incomplete data [14]. These methods are more flexible and agile, allowing for a more comprehensive approach to real-world data [15]. Deep learning algorithms, especially neural networks, such as RNNs, CNNs, and LTSMs, can effectively process temporal data, including signal data from wearable devices. By utilizing the following methods appropriately, deep learning could help automate the detection of diabetes, leading to more effective diagnostics and management of the disease.

This chapter is organized into the following sections. Firstly, in the introduction, the importance and urgency of the observed issue were presented. Subsequently, in the next section, all reviewed methods with the most captivating approaches were collected and presented in a comparison table with respective accuracy for each model. Then, in the discussion part, all observed trends and prospects for the future were described. Lastly, in the conclusion, the most important details were summarized and collected.

8.3 METHODS

The vast majority of reviewed research papers in the area of diabetes mellitus detection using deep learning algorithms used primarily the accuracy metric to evaluate the effectiveness of proposed methods. For preserving consistency in the following comparison, the main measure presented is also the accuracy, which stands for the total count of correctly labeled data.

$$\text{Accuracy} = (\text{TP} + \text{TN})/(\text{TP} + \text{FP} + \text{FN} + \text{TN})$$

where TP – true positive, TN – true negative, FP – false positive, and FN – false negative.

The main advantage of using it as an evaluation metric is easy interpretation and uniformity; however, it may prove to be inefficient in cases with unaddressed class imbalance in the utilized dataset. Otherwise, it provides a valid measure of overall performance (Figure 8.2).

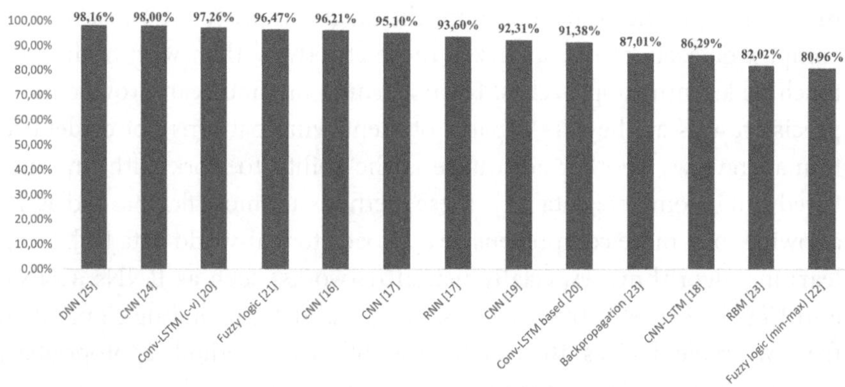

FIGURE 8.2 A graph presenting the accuracy of deep learning methods from the literature review.

In Ref. [16], a medical deep learning model (MDML) was proposed and compared with preexisting methods gaining promising results. The solution introduced by Armghan et al. achieved 96.21% accuracy, 91.53% precision, 94.21% recall, 97.98% F1-score, and 89.90% diagnostic odds ratio [16]. The novelty of the idea is based on providing a tool for analyzing biosensor data for the detection of diabetes based on a combination of convolutional neural networks (CNNs) and recurrent neural networks (RNNs). These sensors are a component of a system that facilitates a remarkably sensitive non-invasive method allowing for the identification of very subtle abnormalities. These biosensors are designed to detect molecules, which are then transformed into electrical signals that are further analyzed in the main system. Obtained data include blood glucose, body temperature, or heart rate.

In Ref. [17], the analysis was based on data derived from electrocardiograms (ECGs) of healthy and affected patients. In the study, both convolutional neural network and CNN-LSTM (long short-term memory) techniques have been implemented and tested with the second one scoring better results. The main features in consideration were heart rate variability (HRV) signals. Swapna et al. presented a hybrid neural network approach, namely fivefold cross-validation CNN-LSTM, achieving 95.1% accuracy. This approach has been tested on 142 datasets in a balanced proportion of both affected and healthy patients. Each dataset contains 1000 samples. After the initial data conversion into heart rate time series, no additional pre-processing has been performed.

In Ref. [18], a one-dimensional CNN (1D-CNN) was proposed by Alex et al. for predicting diabetes mellitus with the highest results of 86.29% accuracy. Additionally, for identifying outliers in missing values, the Tukey method was utilized. For tests and evaluation, a popular dataset was used, namely the Pima Indians Diabetes Dataset [19] from the University of California repository. It includes information about the patient's health and diagnostic measurements, i.e., glucose levels, number of pregnancies the patient has had, BMI, insulin levels, blood pressure, and age [20]. The dataset comprises information about 768 women with 268 patients diagnosed with diabetes and 500 not affected. For dealing with an imbalanced dataset, the synthetic minority oversampling technique (SMOTE) was used. It works by creating new samples for the class present in the minority, resulting in a balanced dataset [21].

In Ref. [22], García-Ordás et al. presented a data augmentation pipeline using an oversampling technique called variational autoencoder (VAE) in combination with sparse autoencoder (SAE) for feature augmentation and CNN. The proposed approach scored 92.31% accuracy. To allow for mutual backpropagation feedback between SAE and CNN, both were trained jointly, which improved the quality of produced features in the process of feature augmentation. In tests, the Pima Indians Diabetes (PID) Database was used. As the amount of features present in the dataset according to the authors was too little to effectively train the neural, more features were generated using the SAE (from 8 to 400).

In Ref. [23], Rahman et al. implemented a Conv-LSTM-based model and a cross-validation Conv-LSTM model, achieving 91.38% accuracy and 97.26% accuracy, respectively. These methods were tested on a widely used PID database. The authors employed the Conv-LSTM, which had not been done earlier. Additionally, to filter out the most important features, a random forest-based Boruta method was applied. This algorithm selects only the most relevant features for later analysis, namely glucose, BMI value, blood pressure, insulin level, and age, while ignoring the rest of the features as those with low significance for the process of classification.

In Ref. [24], a fuzzy rule-based model mixed with the cosine amplitude method was introduced. Aamir et al. implemented and tested these fuzzy approaches on the PID database with 768 records, including data about healthy patients and those affected with diabetes disease. In the evaluation of the presented algorithms, the best one achieved 96.47% accuracy, indicating good prospects for high effectiveness.

In Ref. [25], Kim et al. proposed a robust method consisting of both unsupervised and supervised learning components, that is, fuzzy C-means clustering and a fuzzy min-max neural network learner, respectively. The authors proved their model to outperform the previously known approaches with the detached implementation of these algorithms. In the evaluation of a popular PID database, it achieved 80.96% accuracy. For managing the presence of missing values, researchers used the "fill-in-the-average" mechanism. The highest prevalence of missing values was noticeable for diastolic blood pressure (35), triceps skin fold thickness (221), and 2-hour serum insulin (375) features.

In Ref. [26], both restricted Boltzmann machine and backpropagation were compared and assessed in terms of effectiveness in diagnosing diabetes. Data were subjected to regular data pre-processing, including data splitting and normalization (min-max). In tests with the PID database, the two-layer stochastic neural network (RBM) obtained 82.02% accuracy and the backpropagation algorithm that stems from the multilayer perception (MLP) algorithm achieved 87.01% accuracy.

In Ref. [27], Lee et al. presented an innovative approach to diabetes detection through the use of high-frequency ultrasound and deep learning techniques, namely a CNN-based model that achieved 98% accuracy in efficiency evaluation. The fundamental idea of the authors has its foundations in observed phenomena that glucose tends to gradually build up in the red blood cells forming glycated hemoglobin on their outer layer. The deep learning algorithm was applied to monitor changes in the red blood cells and detect any relations between these and glucose levels and signal data.

In Ref. [28], Nadesh et al. implemented a deep neural network model as an unsupervised approach toward diabetes detection based on the PID database. In data preparation, many techniques were applied such as dimensionality reduction to help mitigate possible overfitting, filter, and wrapper methods, and for determining the most valuable features, extremely randomized trees (feature importance [28]) were used. Feature importance helps to establish the features with the highest significance for later classification by providing a reduction in bias. The proposed method obtained 98.16% accuracy in performance evaluation (Table 8.1).

8.4 DISCUSSION

As the process of detecting and properly diagnosing diabetes mellitus, especially in early stages, may be challenging, it is vital to establish reliable and efficient methods.

The efficacy and benefits of adapting deep learning methods such as CNNs, RNNs, LSTM functionality, and fuzzy logic are prominent. It is noticeable, for example, in the increased interest in deep learning methods application in the medical field, their helpfulness for human experts, and their general high performance, as well as in the growing number of research papers published in this area of concern (Figure 8.3).

The highest performing models are based on hybrid approaches that not only combine and mix different deep learning algorithms but also combine different feature augmentation and pre-processing techniques. Through trial and error, the best combinations can be established and then further improved. These ideas can then be implemented and utilized in practice, here in healthcare to ameliorate the process of diabetes mellitus diagnosis and even enable effective prediabetic state detection.

TABLE 8.1 Comparison of Deep Learning Approaches

References	Year	Model	Techniques	Data	Accuracy (%)
Armghan et al. [16]	2023	Medical deep learning model (MDML)	CNN and RNN	Patient's medical record data and biosensor data (i.e., glucose levels and insulin levels)	96.21
Swapna et al. [17]	2018	Fivefold cross-validation	CNN and CNN-LSTM	Variations of instantaneous heart rate – heart rate variability (HRV) of healthy and affected patients	93.6 95.1
Alex et al. [18]	2022	Deep 1D-convolutional neural network (DCNN)	1D-CNN	Pima Indians Diabetes Dataset (PIDD)	86.29
García-Ordás et al. [22]	2021	CNN classifier, trained jointly with the SAE for featuring augmentation	CNN and sparse autoencoder (SAE)	Prima Indians Diabetes Dataset (PIDD)	92.31
Rahman et al. [23]	2020	Conv-LSTM-based model and cross-validation Conv-LSTM model	Convolutional long short-term memory (Conv-LSTM)	Prima Indians Diabetes Dataset (PIDD)	91.38 97.26
Aamir et al. [24]	2021	Fuzzy rule-based model	Fuzzy logic	Prima Indians Diabetes Dataset (PIDD)	96.47
Kim et al. [25]	2022	A hierarchical combination of fuzzy C-means clustering component (unsupervised) and fuzzy max-min neural network (supervised)	Fuzzy max-min neural network	Prima Indians Diabetes Dataset (PIDD)	80.96
Anggoro et al. [26]	2023	Two-layer stochastic neural network and supervised learning algorithm	Restricted Boltzmann machine (RBM) and backpropagation	Prima Indians Diabetes Dataset (PIDD)	82.02 87.01
Lee et al. [27]	2024	CNN model	CNN	High-frequency ultrasound (HFU) signal data	98.0
Nadesh et al. [28]	2020	Deep neural network	DNN	Prima Indians Diabetes Dataset (PIDD)	98.16

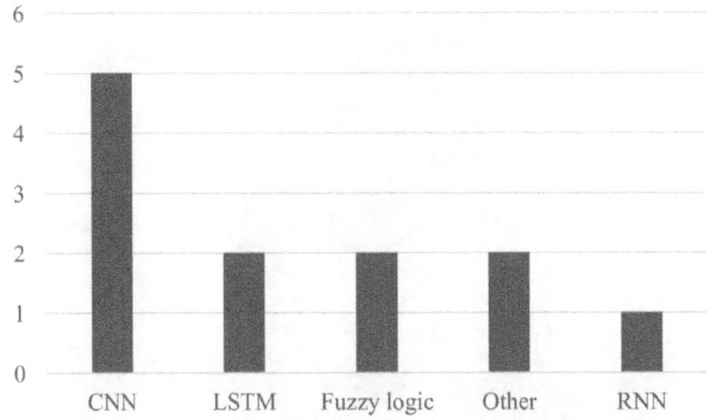

FIGURE 8.3 A graph presenting the prevalence of deep learning methods utilized in the literature review.

From the presented literature review, it is noticeable that the most prevalent deep learning techniques that are widely employed are the CNN algorithm, then hybrid approaches with LSTM, fuzzy logic, and RNNs, and then other less popular approaches such as restricted Boltzmann machine (RBM).

It is important to establish streamlined systems that would allow for effective patient health monitoring, early disease detection, and adroit treatment implementation. According to Ref. [16], one of the preeminent approaches is the utilization of biosensors. These sensors enable precise measurement of the blood glucose levels. Complementing the use of deep learning methods together could form a streamlined real-time monitoring system for early diabetes detection. This particular approach could be particularly beneficial for high-risk patients, i.e., those having a family history of diabetes, having prediabetes, being overweight, having low physical activity, having gestational diabetes, or being 45 years or older [29]. However, the authors argue that the introduced idea may be more applicable for pure detection rather than determining the severity of diabetes or recommending the optimal treatment strategy.

Nevertheless, the proposed solution is a promising way of prompt diagnosis.

Another innovative idea is the approach proposed by Lee et al. [27], where a non-invasive in vivo method without blood collection was implemented. It involved, similarly as in the case of biosensors from another research paper [16], the idea of monitoring signals and data obtained through that examination. This could embody a profound change in the way people affected by diabetes mellitus deal with the disease every day by

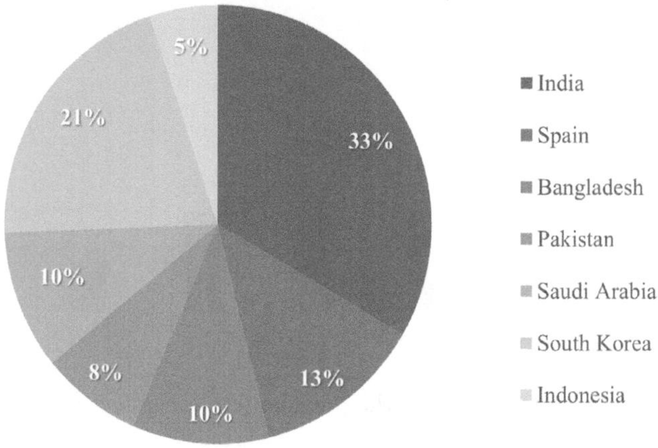

FIGURE 8.4 A pie chart presenting the percentage share of countries in research in deep learning methods for diabetes detection based on the literature review.

allowing them to no longer monitor their current insulin levels through the use of invasive tools and devices.

In terms of deep learning methods applied, another interesting and promising approach is fuzzy logic-based classification, which is relatively easily comprehensible compared to other solutions.

The presented literature review includes works from 2018 to 2024. The number of authors with interest in deep learning methods application in the field of diabetes detection is relatively high. Analyzing the percentage distribution of research papers in terms of geographical origin, it can be noticed that the substantial majority is authored in India and South Korea (Figure 8.4).

8.5 CONCLUSION

Diabetes mellitus is a serious health concern affecting many people worldwide, and this issue needs to be addressed effectively. From the presented literature review, it can be deduced that there is a strong, growing interest in the area of improving the process of diabetes detection and diagnosis with the help of refined deep learning methods that prove to be successful with high-performance accuracy scored in the evaluation. It can be noticed that the methods with the highest accuracy are CNNs, then hybrid approaches such as Conv-LSTM and fuzzy logic, and then RNNs. Nevertheless, all reviewed approaches presented in this review still achieved very high accuracy results demonstrating streamlined functionality and robust classification performance. The countries most interested in the prevailing issue are India and South Korea with the vast majority of papers originating there. There are many innovative ideas for the future of diabetes detection, one being the implementation of non-invasive health state monitoring systems such as leveraging the functionality of biosensors or other signal devices. These approaches could allow for early diabetes detection or prediabetic state detection, which would in fact greatly save patients' health by implementing effective treatment before the disease progresses drastically.

ACKNOWLEDGMENT

The authors would like to thank the Rector of the Silesian University of Technology for providing support on this research project under the IDUB initiative.

REFERENCES

1. Magliano, D. J., Boyko, E. J., & Atlas, I. D. (2021). What is diabetes? In *IDF DIABETES ATLAS [Internet]. 10th edition*. International Diabetes Federation, Brussels, Belgium.
2. Goutham, S., Ravi, V., & Soman, K. P. (2018). Diabetes detection using deep learning algorithms. *ICT Express*, *4*, 243–246. DOI:10.1016/j.icte.2018.10.005.
3. Kaul, K., Tarr, J. M., Ahmad, S. I., Kohner, E. M., & Chibber, R. (2013). Introduction to diabetes mellitus. In Ahmad, S. I. (Ed.), *Diabetes: An Old Disease, a New Insight*, 1–11. New York: Springer.
4. Kawasaki, E. (2014). Type 1 diabetes and autoimmunity. *Clinical Pediatric Endocrinology*, *23*(4), 99–105.
5. Edelman, S. V. (1998). Type II diabetes mellitus. *Advances in Internal Medicine*, *43*, 449–500.
6. Agliata, A., Giordano, D., Bardozzo, F., Bottiglieri, S., Facchiano, A., & Tagliaferri, R. (2023). Machine learning as a support for the diagnosis of type 2 diabetes. *International Journal of Molecular Sciences*, *24*(7), 6775.
7. Rowley, W. R., Bezold, C., Arikan, Y., Byrne, E., & Krohe, S. (2017). Diabetes 2030: insights from yesterday, today, and future trends. *Population Health Management*, *20*(1), 6–12.
8. Fregoso-Aparicio, L., Noguez, J., Montesinos, L., & García-García, J. A. (2021). Machine learning and deep learning predictive models for type 2 diabetes: a systematic review. *Diabetology & Metabolic Syndrome*, *13*(1), 1–22.
9. Megari, K. (2013). Quality of life in chronic disease patients. *Health Psychology Research*, 1(3), e27.
10. WHO. *The Global Health Observatory*. https://www.who.int/data/gho/indicator-metadata-registry/imr-details/2380 (access date: 19.02.2024).

11. Zou, Q., Qu, K., Luo, Y., Yin, D., Ju, Y., & Tang, H. (2018). Predicting diabetes mellitus with machine learning techniques. *Frontiers in Genetics, 9*, 515.
12. Swapna, G., Vinayakumar, R., & Soman, K. P. (2018). Diabetes detection using deep learning algorithms. *ICT Express, 4*(4), 243–246.
13. Rawat, V., Joshi, S., Gupta, S., Singh, D. P., & Singh, N. (2022). Machine learning algorithms for early diagnosis of diabetes mellitus: a comparative study. *Materials Today: Proceedings, 56*, 502–506.
14. Zhu, T., Li, K., Herrero, P., & Georgiou, P. (2020). Deep learning for diabetes: a systematic review. *IEEE Journal of Biomedical and Health Informatics, 25*(7), 2744–2757.
15. Tsiknakis, N., Theodoropoulos, D., Manikis, G., Ktistakis, E., Boutsora, O., Berto, A., … & Marias, K. (2021). Deep learning for diabetic retinopathy detection and classification based on fundus images: a review. *Computers in Biology and Medicine, 135*, 104599.
16. Armghan, A., Logeshwaran, J., Sutharshan, S. M., Aliqab, K., Alsharari, M., & Patel, S. K. (2023). Design of biosensor for synchronized identification of diabetes using deep learning. *Results in Engineering, 20*, 101382.
17. Swapna, G., Soman, K. P., & Vinayakumar, R. (2018). Automated detection of diabetes using CNN and CNN-LSTM network and heart rate signals. *Procedia computer science, 132*, 1253–1262.
18. Alex, S. A., Nayahi, J. J. V., Shine, H., & Gopirekha, V. (2022). Deep convolutional neural network for diabetes mellitus prediction. *Neural Computing and Applications, 34*(2), 1319–1327.
19. Blake, C. L., & Merz, C. J. (1998) *UCI Repository of Machine Learning Databases*. Department of Information and Computer Science, University of California, Irvine, CA.
20. Kaggle. *Pima Indians Diabetes Database*. https://www.kaggle.com/datasets/uciml/pima-indians-diabetes-database (access date: 07.03.2024).
21. Maniruzzaman, M., Kumar, N., Islam, S., Suri, H. S., El-Baz, A. S., Suri, J. S., & Abedin, M. M. (2017). Comparative approaches for classification of diabetes mellitus data: machine learning paradigm. *Computer Methods and Programs in Biomedicine, 152*, 23–34.
22. García-Ordás, M. T., Benavides, C., Benítez-Andrades, J. A., Alaiz-Moretón, H., & García-Rodríguez, I. (2021). Diabetes detection using deep learning techniques with oversampling and feature augmentation. *Computer Methods and Programs in Biomedicine, 202*, 105968.
23. Rahman, M., Islam, D., Mukti, R. J., & Saha, I. (2020). A deep learning approach based on convolutional LSTM for detecting diabetes. *Computational Biology and Chemistry, 88*, 107329.
24. Aamir, K. M., Sarfraz, L., Ramzan, M., Bilal, M., Shafi, J., & Attique, M. (2021). A fuzzy rule-based system for classification of diabetes. *Sensors, 21*(23), 8095.
25. Kim, K. B., Park, H. J., & Song, D. H. (2022). Combining supervised and unsupervised fuzzy learning algorithms for robust diabetes diagnosis. *Applied Sciences, 13*(1), 351.

26. Anggoro, D. A., & Hajiati, S. (2023). Comparison of performance metrics level of restricted Boltzmann machine and backpropagation algorithms in detecting diabetes mellitus disease. *Iraqi Journal of Science*, *64*, 907–921.
27. Lee, J. E., Jeon, H. J., Lee, O. J., & Lim, H. G. (2024). Diagnosis of diabetes mellitus using high frequency ultrasound and convolutional neural network. *Ultrasonics*, *136*, 107167.
28. Nadesh, R. K., & Arivuselvan, K. (2020). Type 2: diabetes mellitus prediction using deep neural networks classifier. *International Journal of Cognitive Computing in Engineering*, *1*, 55–61.
29. Centers for Disease Control and Prevention. *Diabetes Risk Factors*. https://www.cdc.gov/diabetes/basics/risk-factors.html (assess date: 07.03.2024).

CHAPTER 9

Examining the Role of Machine Learning and Deep Learning in Diabetes Mellitus Detection and Diagnosis

A Critical Review

C. Gunavathi, Akshat Bokdia, and Siri R. Kulakarni

9.1 INTRODUCTION

The detection and diagnosis of diabetes mellitus (DM) have evolved significantly, incorporating both traditional and modern methods. Traditional approaches, such as blood glucose monitoring and clinical tests, remain fundamental. However, advancements in ML and DL have revolutionized the field. Techniques like ensemble methods and different classifier algorithms enhance data analysis. Meanwhile, various DL architectures are also used for tasks such as diabetic retinopathy detection from retinal

DOI: 10.1201/9781032708430-9

images. This integration of traditional and advanced methods promises more accurate and efficient diabetes diagnosis and management.

Many research papers employ various ML techniques for diabetes detection and diagnosis. These techniques include classification methods such as support vector machine (SVM), decision tree, logistic regression, *k*-nearest neighbors (KNN), and random forest. Ensemble methods such as AdaBoost, gradient boosting, and bagging are utilized to improve the model performance through the combination of multiple base learners. SVMs are employed for classification tasks, leveraging their capacity to effectively handle high-dimensional data and nonlinear relationships. Decision trees are utilized for their interpretability and ability to handle categorical data, while logistic regression is applied for binary classification tasks. KNN algorithms are utilized for their simplicity and effectiveness in handling classification tasks based on similarity measures. Random forest algorithms are employed for their robustness and ability to handle high-dimensional data with complex relationships.

DL techniques play a significant role in diabetes detection and diagnosis. Many existing works utilize various DL architectures, including convolutional neural networks (CNNs), recurrent neural networks (RNNs), long short-term memory networks (LSTMs), and autoencoders. CNNs are particularly effective for image-based data such as retinal images, leveraging their ability to automatically learn hierarchical features. RNNs and LSTMs are applied for sequential data such as time-series signals, enabling the modeling of temporal dependencies. Autoencoders are utilized for feature learning and dimensionality reduction, which is useful in feature extraction from high-dimensional data. Transfer learning techniques are also employed, leveraging pre-trained deep learning (DL) models on large datasets to improve the performance on smaller datasets.

Several papers specifically focus on CNN-based approaches for diabetes detection and diagnosis. These papers utilize CNN architectures such as AlexNet, VGG, ResNet, and Inception, which have demonstrated good results in image classification processes. Transfer learning techniques are commonly employed, where pre-trained CNN models trained on large-scale image datasets (e.g., ImageNet) are fine-tuned on diabetic retinopathy datasets to improve classification accuracy. Data augmentation techniques such as rotation, scaling, and flipping are applied to increase the diversity of training samples and improve model generalization. Additionally, attention mechanisms and ensemble learning techniques are utilized to further enhance the performance of CNN-based models for diabetic retinopathy detection.

The papers used for the analysis collectively focus on DM detection, diagnosis, and prediction using various ML and DL techniques. Methods include utilizing diverse data modalities such as clinical data, retinal images, DNA sequences, photoplethysmography signals, foot thermography, and heart rate variability signals. There is a strong emphasis on the application of ML and DL algorithms for enhancing diabetes diagnosis and prediction accuracy. Techniques include ensemble methods, hybrid models, feature selection techniques, and augmentation methods to improve model performance. Papers explore the potential for early detection and real-time monitoring of diabetes using wearable sensors, biosensors, and electronic noses. The goal is to enable timely intervention, particularly in critical care settings, to mitigate the adverse effects of diabetes.

9.2 METHODOLOGY

This section outlines the methodology employed to identify relevant research works related to diabetes detection and prediction, including the formulation of research questions, search strategy, and review results.

9.2.1 Research Questions

The study aims to address the following research questions:

- How are ML and DL techniques utilized in diabetes detection and prediction, and what are their implications for clinical practice?
- What factors influence the accuracy and effectiveness of diabetes prediction models?
- Which ML algorithms and DL architectures demonstrate superior performance in diabetes detection and prediction?
- What datasets are commonly used in training and evaluating diabetes prediction models, and how do they impact model performance?
- What is the current state of ML and DL approaches in providing accurate prognoses for diabetic patients?

9.2.2 Search Strategy

The search for relevant research articles was conducted using Litmaps, an advanced literature mapping tool. Various academic repositories, including PubMed, IEEE, and Science Direct, were systematically queried using relevant keywords such as diabetes/machine learning/deep learning/

prediction. The search aimed to retrieve articles that investigate diabetes detection and prediction using ML or DL techniques.

9.2.3 Review Results

Following a rigorous selection process to ensure relevance to the study objectives, the initial pool of articles is narrowed down to 41 papers. The selection process involves screening for relevance based on title, abstract, and full-text examination. The final set of articles represents a comprehensive review of the current literature on diabetes detection and prediction using machine learning (ML) and DL approaches.

9.2.4 Study Analysis

After the title and abstract were reviewed, the articles were removed, out of a total of 80, for a variety of reasons.

The following factors led to the articles' exclusion:

i. Text accessibility is lacking.

ii. Lack of DL or ML models.

iii. Exclusive attention to abstract ideas.

iv. Examine scholarly works.

v. Just posters or abstracts.

9.2.5 Dataset

In the domain of DM detection and prediction, various datasets serve as the foundation for developing and evaluating ML and DL models. These datasets encompass diverse parameters and modalities adapted to the individual objectives of each study. For instance, the Pima Indian Diabetes Database (PIDD) is frequently utilized. Other datasets include photoplethysmography signals for hypertension risk stratification, foot thermography images, and DNA sequences for DL-based diagnosis. Additionally, datasets sourced from preventive healthcare examinations, retinal images, electrocardiogram (ECG) signals, and clinical registries further contribute to the multifaceted landscape of diabetes detection and prediction research. These datasets provide researchers with invaluable resources for training and evaluating predictive models, paving the way for advancements in diabetes management and early intervention strategies.

9.2.6 Data Preprocessing

Data preprocessing is a critical step in the data analysis pipeline aimed at cleaning and transforming raw data into a usable format for ML models. It involves tasks such as handling missing values, normalization, feature scaling, and encoding categorical variables. By preprocessing the data, analysts can enhance model performance and reduce the likelihood of errors caused by inconsistent or incomplete data. Table 9.1 summarizes the data preprocessing techniques used in the reviewed articles.

Feature selection is a process of identifying and selecting the most relevant features from a dataset to improve model performance and reduce overfitting. This involves techniques such as univariate selection, recursive feature elimination, and feature importance scoring. By selecting only the

TABLE 9.1 Data Preprocessing Techniques

Preprocessing Techniques	References
Fuzzy logic-based preprocessing	Priyadarshini and Lakshmi Shrinivasan (2021)
Chebyshev type-II band-pass filtering, cubic spline estimation for baseline correction, and Savitzky-Golay smoothing filter	Singh and Nirala (2023)
Singular value decomposition (SVD) to predict missing values in the dataset	Nilashi et al. (2023)
Min-max normalization	Kamble and Patil (2016) and García-Ordás et al. (2021)
Cleaning the raw data collected from biosensors by removing outliers and noise, and then standardizing the data through normalization	Armghan et al. (2023)
Standard techniques such as normalization and scaling	Rahman et al. (2020)
Outlier detection and removal and imputation of missing values using the multiple imputation by chained equations (MICE) method	Kopitar et al. (2020) and Chowdary and Kumar (2021)
Multiple imputations were used to handle missing data, followed by scaling continuous variables and transforming categorical variables into dummy variables	Karmand et al. (2024)
Data cleansing, handling missing values through mean imputation, and one-hot encoding for categorical variables	Utomo et al. (2024)

(Continued)

TABLE 9.1 (*Continued*) Data Preprocessing Techniques

Preprocessing Techniques	References
Data augmentation and imputation using generative adversarial networks	Boughareb et al. (2023)
Discrete wavelet transform (DWT) and Z-score normalization	Gudiño-Ochoa et al. (2024)
Data normalization	Chee et al. (2024)
Z-score normalization, followed by outlier detection using the interquartile range (IQR) method and subsequent replacement with the median	Reza et al. (2024)
Data imputation and normalization	Lu et al. (2024)
Fifth-order Butterworth filter, outlier detection, and data normalization	Kim et al. (2024)
KNN imputer	Alnowaiser (2024)
Mean and mode imputation	Noviyanti and Alamsyah (2024)

most informative features, feature selection helps simplify models, decrease computational complexity, and enhance interpretability. Table 9.2 summarizes all the feature selection techniques used in the reviewed articles.

9.2.7 Techniques Used in the Study

The premise underlying the methodology in the papers on diabetes detection and diagnosis is to leverage ML and DL algorithms alongside diverse data modalities, such as clinical data and imaging techniques, to enhance

TABLE 9.2 Feature Selection Techniques

Feature Selection Techniques	References
STEPDISC, a step-by-step discriminant analysis to select a subset of quantitative variables for class discrimination	Nilashi et al. (2023)
Hybrid method combining filter and wrapper methods	Armghan et al. (2023)
Boruta algorithm based on random forests	Rahman et al. (2020)
Filtering out variables with over 50% missing values and selecting relevant predictor variables related to physiological and lifestyle factors.	Kopitar et al. (2020)
Random forest algorithm	Chowdary and Kumar (2021)
Recursive feature elimination (RFE) integrated with a tree-based ML model	Karmand et al. (2024)
Univariate feature selection	Gudiño-Ochoa et al. (2024)
Variables were chosen based on existing literature, data completeness, and evaluation of collinearity	Lu et al. (2024)

the accuracy and efficiency of early detection, diagnosis, and prediction of diabetes. By harnessing advanced computational techniques and integrating various medical imaging modalities and genetic data, the goal is to facilitate timely intervention and personalized treatment strategies, ultimately improving health outcomes for individuals with diabetes. Tables 9.3–9.6 gives the complete summary of the ML, DL, and CNN techniques used in the existing research papers.

9.3 RESEARCH GAP

9.3.1 Dataset Size and Diversity

Many studies in the field of diabetes detection and diagnosis highlight a significant limitation stemming from the use of small or single-source datasets. This limitation undermines the generalizability and reliability of the findings, as they may not adequately represent the diversity of patient populations or clinical scenarios. To address this gap, future research efforts should prioritize validation on larger and more diverse datasets, encompassing a wider range of demographic factors, clinical variables, and geographic regions.

9.3.2 Algorithmic Exploration and Optimization

A common research gap identified across various papers is the need for further exploration of DL architectures, feature selection methods, and optimization techniques. While many studies achieve promising results, there remains uncertainty regarding the scalability and applicability of proposed models beyond specific datasets or demographics. Future research endeavors should focus on enhancing algorithmic efficiency, optimizing model parameters, and assessing performance across diverse populations to ensure robust and reliable diabetes detection and diagnosis systems.

9.3.3 Clinical Interpretability and Real-World Deployment

Despite advancements in artificial intelligence (AI)-based diabetes detection methods, there is a persistent gap in addressing the interpretability of models and their suitability for real-world clinical settings. Many studies lack discussion on how the proposed algorithms can be integrated into existing clinical workflows or effectively utilized by healthcare practitioners. To bridge this gap, future research should prioritize the development of interpretable AI models and explore strategies for seamless integration with clinical decision support systems, ultimately enhancing their utility in healthcare practice.

TABLE 9.3 Machine Learning Techniques for Diabetes Mellitus Detection

References	Methodology	Dataset	Performance Technique	Metric	Value
Nilashi et al. (2023)	Singular value decomposition (SVD), self-organizing map (SOM) clustering, and deep belief network (DBN) ensembles	Pima Indians diabetes database (PIDD)	Detection	Accuracy	98.32%
Kamble and Patil (2016)	Restricted Boltzmann machine	Not explicitly mentioned	Detection	Accuracy	80.0%
Kopitar et al. (2020)	XGBoost, Glmnet	EHR data	Detection	RMSE	0.881%, 0.859%
Karmand et al. (2024)	Gradient boosting machine (GBM) model	Baseline data from the Fasa Adult Cohort Study (FACS)	Detection	Accuracy, precision, sensitivity, specificity, F1-score, AUC	0.92%, 0.53%, 0.24%, 0.98%, 0.33%, 0.75%
Utomo et al. (2024)	Machine learning model based on the CatBoost algorithm	Medical data from MIT's GOSSIS initiative	Prediction	Accuracy	83.7%
Gudiño-Ochoa et al. (2024)	XGBoost DNN	Own data collection	Detection	Accuracy	95% 94.4%
Alnowaiser (2024)	Tri-ensemble of XGBoost, random forest, and extra trees classifiers	Pima Indians Diabetes Database (PIDD)	Prediction	Accuracy, precision, recall, F1-score	97.49%, 98.16%, 99.35%, 98.84%
Noviyanti and Alamsyah (2024)	Random forest model	Pima Indians Diabetes Database (PIDD)	Detection	Accuracy	87%
Olisah et al. (2022)	Random forest, SVM, 2GDNN	PIMA Indian Dataset and the Laboratory of the Medical City Hospital (LMCH) diabetes dataset	Prediction	Accuracy, precision, sensitivity, F1-score	97.33%, 97.28%, 97.33%, 97.27%

TABLE 9.4 Deep Learning Techniques for Diabetes Mellitus Detection

References	Methodology	Dataset	Performance Technique	Metric	Value
Priyadarshini and Shrinivasan (2021)	ANFIS & fuzzy logic integration	Pima Indian Diabetes Database (PIDD)	Classification	Accuracy, sensitivity, specificity	86.48%, 92.78%, 73.08%
Armghan et al. (2023)	High-sensitive biosensors coupled with medical DNN	Not explicitly mentioned	Detection	Accuracy, precision, recall, F1-score	96.21%, 91.53%, 94.21%, 97.98%
Boughareb et al. (2023)	Hybrid GAN-ANN	Pima Indians Diabetes Database (PIDD)	Prediction	Accuracy	94%
Gudiño-Ochoa et al. (2024)	XGBoost DNN	Own data collection	Detection	Accuracy	95%, 94.4%
Reza et al. (2024)	Stacking ensemble with three neural network architectures	Pima Indians Diabetes Database (PIDD)	Classification	Accuracy, precision, recall, F1-score	95.5%, 94%, 97%, 96%
Lu et al. (2024)	DNN model	Data from the Canadian Primary Care Sentinel Surveillance Network (CPCSSN)	Detection	Accuracy	71%
Kim et al. (2024)	Deep learning models trained on ECG signals	Sourced from SNUH-HSGC and SNUH-HPC	Detection	AUROC	0.816%

(*Continued*)

TABLE 9.4 (*Continued*) Deep Learning Techniques for Diabetes Mellitus Detection

References	Methodology	Dataset	Performance Technique	Metric	Value
Nadesh and Arivuselvan (2022)	DNN	Pima Indian Diabetes Dataset (PIDD)	Classification	Accuracy	98.16%
Riyad et al. (2020)	Deepnet	Datasets acquired from the Ministry of National Guard Hospital Affairs (MNGHA) in Saudi Arabia between 2013 and 2015	Prediction	Accuracy, precision, recall, F-Measure, PhiCoefficient	88.48%, 88.29%, 88.36%, 0.88%, 0.77%
Nagaraj and Deepalakshmi (2021)	ESVM, DNN	PIMA Indian dataset	Prediction	Accuracy, sensitivity, specificity	98%, 99.43%, 95.70%
Suma et al. (2022)	Deep learning-based object detection algorithm	Nailfold video capillaroscopy performed on 120 adult patients	Detection	Accuracy, sensitivity, specificity	88.22%, 89%, 89%

TABLE 9.5 Convolutional Neural Networks for Diabetes Mellitus Detection

References	Methodology	Dataset	Performance Technique	Parameter	Values Obtained
Islam et al. (2021)	DiaNet model (multi-stage CNN)	QBB retinal image dataset	Prediction	Accuracy, precision, recall, specificity, F1-score, AUC ROC	84.47%, 83.59%, 85.86%, 83.06%, 84.71%, 84.46%
Swapna et al. (2018)	CNN and CNN+LSTM	ECG data from 20 diabetes patients and 20 normal individuals, with heart rate variability signals derived using Pan and Tompkinson's algorithm	Detection	Accuracy (CNN+LSTM), Accuracy (CNN)	95.91%, 93.6%
Aslan and Sabanci (2023)	ResNet18 and ResNet50 CNN models, SVM	PIMA dataset	Prediction	Accuracy, precision, recall, F1-score, MCC	92.19%, 92.97%, 95.20%, 94.07%, 82.68%
Gayathri and Varun (2020)	CNN	Generic IDRiD, MESSIDOR, and KAGGLE datasets	Classification	Accuracy	99.89%
Aswiga et al. (2023)	CNN	150 samples, encompassing 38 features related to diabetes, including parameters like age, Alb/Crea Ratio, SGOT, SGPT, etc.	Detection	Accuracy	-
Madan et al. (2022)	CNN and Bi – directional LSTM	PIMA Indian Diabetes Database (PIDD)	Prediction	Accuracy, sensitivity, specificity	98%, 97%, 98%
Nahiduzzaman et al. (2021)	CNN-SVD	APTOS-2019 blindness detection and Messidor-2 datasets	Feature extraction and reduction	Accuracy, AUC, F1-score, recall, precision	98.09%, 99.41%, 97%, 96%, 97%

(Continued)

TABLE 9.5 (*Continued*) Convolutional Neural Networks for Diabetes Mellitus Detection

References	Methodology	Dataset	Performance Technique	Parameter	Values Obtained
Dayal et al. (2020)	CNN	—	Detection	—	—
Bhopte and Rai (2022)	CNN + LSTM	Pima Indian Diabetes Dataset (PIDD)	Prediction	Accuracy, precision, recall, F1-score	89.30%, 87.80%, 84.10%, 85.58%
Macsik et al. (2022)	Local binary CNN	APTOS and EyePACS	Classification	Accuracy, AUC, specificity, sensitivity, NPV, PPV	96.58%, 0.9871, 96.59%, 94.35%, 94.35%, 96.73%
Zanelli et al. (2023)	CNN	DB_DT2: Type 2 diabetic PPG waves, DB_shape: Photoplethysmography (PPG) waves classified based on quality and shape, DB_HT: Hypertension (HT) PPG waves.	Detection	—	—

TABLE 9.6 Convolutional Neural Networks for Diabetes Mellitus Detection

References	Methodology	Dataset	Performance Technique	Parameter	Values Obtained
El-Attar et al. (2022)	CNN+LSTM	Human insulin dataset from GenBank	Classification	Accuracy, precision, sensitivity, specificity, F1-score	99%, 98%, 100%, 98%, 99%
Lahmar and Idri (2022)	CNN and machine learning	APTOS	Classification	Accuracy	96.22%
Albahli and Yar (2022)	CNN	Indian Diabetic Retinopathy Image Dataset (IDRiD)	Detection	Accuracy	89.69%
Ayat et al. (2024)	CNN+LSTM	Pima Dataset	Classification	Accuracy	97%
Amalia et al. (2021)	CNN+LSTM	MESSIDOR	Detection	Accuracy	90.22%
Chee et al. (2024)	CNN-LSTM using gait acceleration	Obtained from Sánchez-DelaCruz et al. (2019)	Detection	Accuracy	91.25%
Chowdary and Kumar (2021)	CLSTM LSTM	Pima Indians Diabetes Database (PIDD)		Accuracy	96.8% 92.5%
Rahman et al. (2020)	Convolutional long short-term memory	Pima Indians Diabetes Database (PIDD)	Detection	Accuracy	97.26%
Anaya-Isaza and Zequera-Diaz (2022)	ResNet50v2 network	Plantar Thermogram Database for the Study of Diabetic Foot Complications	Detection	Accuracy	100%
García-Ordás et al. (2021)	CNN	Pima Indians Diabetes Database (PIDD)	Detection	Accuracy	92.31%

9.3.4 Feature Engineering and Preprocessing Techniques

Several papers in the field of diabetes detection highlight the importance of sophisticated feature engineering and preprocessing techniques to improve model performance and address data-related challenges. However, there remains a gap in understanding the interpretability and potential biases introduced by these techniques. Future research endeavors should prioritize the exploration of transparent feature engineering methods and rigorous preprocessing pipelines to ensure the reliability and fairness of AI-based diabetes detection systems.

9.3.5 Model Validation and Generalization

One of the critical challenges in AI-based diabetes detection research is the validation of proposed models on external datasets and their generalizability across diverse populations. Many studies focus on achieving high accuracy on specific datasets without adequately assessing the robustness of the models in real-world clinical settings. To address this gap, future research efforts should prioritize validation on diverse and representative datasets, encompassing variations in patient demographics, disease severity, and clinical protocols.

9.3.6 Ethical and Bias Considerations

Despite the potential benefits of AI-based diabetes detection systems, there are significant ethical considerations related to bias, fairness, and transparency that remain largely unaddressed in current research. Many studies overlook potential biases introduced by dataset characteristics and algorithmic approaches, raising concerns about the reliability and equity of AI-based healthcare solutions. To mitigate these risks, future research endeavors should prioritize fairness-aware AI techniques, transparent reporting practices, and interdisciplinary collaboration to ensure the ethical deployment of diabetes detection systems in clinical practice.

9.3.7 Application-Specific Challenges

Research focusing on specific applications, such as retinopathy detection or biosensor development, highlights unique challenges that require tailored solutions. For example, image quality, sensor sensitivity, and model interpretability are critical considerations in retinopathy detection, while biosensor development may require addressing challenges related to sensor accuracy and real-time data processing. To overcome these application-specific challenges, future research efforts should prioritize interdisciplinary collaboration and domain-specific expertise, ensuring the development of robust and effective diabetes detection solutions tailored to specific clinical contexts.

9.3.8 Model Complexity and Scalability

Many AI-based diabetes detection models proposed in the literature exhibit high complexity, raising concerns about their scalability and computational efficiency, particularly in real-time or resource-constrained environments. While complex models may achieve high accuracy on benchmark datasets, their practical utility in clinical settings may be limited by computational constraints and deployment challenges. To address this gap, future research should explore lightweight model architectures, optimization techniques, and scalable deployment strategies to ensure the feasibility and effectiveness of AI-based diabetes detection systems in real-world healthcare environments.

9.3.9 Validation and Interpretation of Results

Despite the emphasis on achieving high accuracy in diabetes detection, many studies lack qualitative analysis supporting model outputs and fail to provide insights into the clinical significance of their findings. This gap undermines the interpretability and practical utility of research outcomes, hindering their adoption in clinical practice. To address this gap, future research efforts should prioritize transparent reporting practices, qualitative analysis of model outputs, and collaboration with healthcare professionals to ensure the relevance and applicability of AI-based diabetes detection systems in real-world clinical settings.

9.3.10 Data Availability and Usability

A recurring challenge in AI-based diabetes detection research is the limited availability and usability of healthcare datasets. Many studies rely on proprietary or restricted datasets, hindering collaboration and reproducibility in the research community. To address this gap, future research efforts should prioritize data-sharing initiatives, open-access repositories, and standardized evaluation frameworks to facilitate the development and validation of robust and generalizable AI-based diabetes detection systems. By addressing these challenges, researchers can accelerate progress toward developing clinically impactful solutions for diabetes detection and diagnosis.

9.4 CONCLUSION

In conclusion, the field of DM prediction and detection has seen significant advancements, blending traditional methods like blood glucose monitoring with cutting-edge ML and DL techniques. By integrating ensemble methods, support vector machines, decision trees, and DL architectures such as CNNs and RNNs, we can achieve more accurate diagnosis and early detection of diabetic complications. This fusion of traditional and modern approaches promises improved patient outcomes and underscores the importance of continued research and collaboration in diabetes management.

REFERENCES

Albahli, S., and G. N. A. H. Yar. "Detection of diabetic retinopathy using custom CNN to segment the lesions." *Intelligent Automation & Soft Computing* 33, no. 2 (2022): 837–853.

Alnowaiser, K. "Improving healthcare prediction of diabetic patients using KNN imputed features and tri-ensemble model." *IEEE Access* 12 (2024): 16783–16793.

Alshammari, R., N. Atiyah, T. Daghistani, and A. Alshammari. "Improving accuracy for diabetes mellitus prediction by using deepnet." *Online Journal of Public Health Informatics* 12, no. 1 (2020): e11.

Amalia , R., A. Bustamam, and D. Sarwinda. "Detection and description generation of diabetic retinopathy using convolutional neural network and long short-term memory." Journal of Physics: Conference Series 1722, no. 1 (2021): 012010. IOP Publishing.

Anaya -Isaza, A., and M. Zequera-Diaz. "Detection of diabetes mellitus with deep learning and data augmentation techniques on foot thermography." *IEEE Access* 10 (2022): 59564–59591.

Armghan, A., J. Logeshwaran, S. M. Sutharshan, K. Aliqab, M. Alsharari, and S. K. Patel. "Design of biosensor for synchronized identification of diabetes using deep learning." *Results in Engineering* 20 (2023): 101382.

Aslan, M. F., and K. Sabanci. "A novel proposal for deep learning-based diabetes prediction: converting clinical data to image data." *Diagnostics* 13, no. 4 (2023): 796.

Aswiga, R. V., M. Karpagam, M. Chandralekha, and S. Deena. "Automated detection and classifying diabetes mellitus using CNN." *Soft Computing: A Fusion of Foundations, Methodologies and Applications* 27 (2023): 6869–6875.

Ayat, Y., W. Benzekri, A. El Moussati, I. Mir, M. Benzaouia, and A. El Aouni. "Novel diabetes classification approach based on CNN-LSTM: Enhanced performance and accuracy." *Diagnostyka* 25 (2024). https://doi.org/10.29354/diag/183633.

Bhopte, M., and M. Rai. "Hybrid deep learning CNN-LSTM model for diabetes prediction." *International Journal of Scientific Research & Engineering Trends* 8 (2022): 444–447.

Boughareb, D., H. Bensalah, and H. Seridi. "A hybrid GAN-ANN-based model for diabetes prediction." *International Journal of Scientific Research in Science and Technology* 10, no. 14 (2023): 30–41.

Chee, L. Z., S. Sivakumar, K. H. Lim, and A. A. Gopalai. "Gait acceleration-based diabetes detection using hybrid deep learning." *Biomedical Signal Processing and Control* 92 (2024): 105998.

Chowdary, P. B. K., and R. U. Kumar. "An effective approach for detecting diabetes using deep learning techniques based on convolutional LSTM networks." *International Journal of Advanced Computer Science and Applications* 12, no. 4 (2021). https://doi.org/10.14569/IJACSA.2021.0120466.

Dayal, K. V., P. Singh, N. Sharma, and K. B. Dubey. "Detection of diabetic retinopathy using CNN." *International Research Journal of Engineering and Technology* 7 (2020): 1131–1134.

El-Attar, N. E., B. M. Moustafa, and W. A. Awad. "Deep learning model to detect diabetes mellitus based on DNA sequence." *Intelligent Automation & Soft Computing* 31, no. 1 (2022): 1–10.

García-Ordás, M. T., C. Benavides, J. A. Benítez-Andrades, H. Alaiz-Moretón, and I. García-Rodríguez. "Diabetes detection using deep learning techniques with oversampling and feature augmentation." *Computer Methods and Programs in Biomedicine* 202 (2021): 105968.

Gayathri, S., V. P. Gopi, and P. Palanisamy. "A lightweight CNN for diabetic retinopathy classification from fundus images." *Biomedical Signal Processing and Control* 62 (2020): 102115.

Gudiño-Ochoa, A., J. A. García-Rodríguez, R. Ochoa-Ornelas, J. I. Cuevas-Chávez, and D. A. Sánchez-Arias. "Noninvasive diabetes detection through human breath using TinyML-powered e-nose." *Sensors* 24, no. 4 (2024): 1294.

Islam, M. T., H. R. H. Al-Absi, E. A. Ruagh, and T. Alam. "DiaNet: A deep learning based architecture to diagnose diabetes using retinal images only." *IEEE Access* 9 (2021): 15686–15695.

Kamble, T. P., and S. T. Patil. "Diabetes detection using deep learning approach." *International Journal for Innovative Research in Science & Technology* 2, no. 12 (2016): 342–349.

Karmand, H., A. Andishgar, R. Tabrizi, A. Sadeghi, B. Pezeshki, M. Ravankhah, E. Taherifard, and F. Ahmadizar. "Machine-learning algorithms in screening for type 2 diabetes mellitus: Data from Fasa Adults Cohort Study." *Endocrinology, Diabetes & Metabolism* 7, no. 2 (2024): e00472.

Kim, J., H.-L. Yang, S. H. Kim, S. Kim, J. Lee, J. Ryu, K. Kim et al. "Deep learning-based long-term risk evaluation of incident type 2 diabetes using electrocardiogram in a non-diabetic population: a retrospective, multicentre study." *EClinicalMedicine* 68 (2024): 102445.

Kopitar, L., P. Kocbek, L. Cilar, A. Sheikh, and G. Stiglic. "Early detection of type 2 diabetes mellitus using machine learning-based prediction models." *Scientific Reports* 10, no. 1 (2020): 11981.

Lahmar, C., and A. Idri. "Classifying diabetic retinopathy using CNN and machine learning." *Bioimaging* 2 (2022): 52–62.

Lu, K., P. Sheth, Z. L. Zhou, K. Kazari, A. Guergachi, K. Keshavjee, M. Noaeen, and Z. Shakeri. "Identifying prediabetes in canadian populations using machine learning." *medRxiv* (2024): 2024-02.

Macsik, P., J. Pavlovicova, J. Goga, and S. Kajan. "Local binary CNN for diabetic retinopathy classification on fundus images." *Acta Polytechnica Hungarica* 19, no. 7 (2022): 27–45.

Madan, P., V. Singh, V. Chaudhari, Y. Albagory, A. Dumka, R. Singh, A. Gehlot, M. Rashid, S. S. Alshamrani, and A. S. AlGhamdi. "An optimization-based diabetes prediction model using CNN and bi-directional LSTM in real-time environment." *Applied Sciences* 12, no. 8 (2022): 3989.

Nadesh, R. K., and K. Arivuselvan. "Type 2: diabetes mellitus prediction using deep neural networks classifier." *International Journal of Cognitive Computing in Engineering* 1 (2020): 55–61.

Nagaraj, P., and P. Deepalakshmi. "Diabetes prediction using enhanced SVM and deep neural network learning techniques: An algorithmic approach for early screening of diabetes." *International Journal of Healthcare Information Systems and Informatics (IJHISI)* 16, no. 4 (2021): 1–20.

Nahiduzzaman, M., M. Robiul Islam, S. M. Riazul Islam, M. O. F. Goni, M. S. Anower, and K.-S. Kwak. "Hybrid CNN-SVD based prominent feature extraction and selection for grading diabetic retinopathy using extreme learning machine algorithm." *IEEE Access* 9 (2021): 152261–152274.

Nilashi, M., R. A. Abumalloh, S. Alyami, A. Alghamdi, and M. Alrizq. "A combined method for diabetes mellitus diagnosis using deep learning, singular value decomposition, and self-organizing map approaches." *Diagnostics* 13, no. 10 (2023): 1821.

Noviyanti, C. N., and A. Alamsyah. "Early detection of diabetes using random forest algorithm." *Journal of Information System Exploration and Research* 2, no. 1: (2024). https://doi.org/10.52465/joiser.v2i1.245.

Olisah, C. C., L. Smith, and M. Smith. "Diabetes mellitus prediction and diagnosis from a data preprocessing and machine learning perspective." *Computer Methods and Programs in Biomedicine* 220 (2022): 106773.

Priyadarshini, L., and L. Shrinivasan. "Diabetes Mellitus Detection and Diagnosis Using AI Classifier." In *Cybernetics, Cognition and Machine Learning Applications: Proceedings of ICCCMLA 2020*, August, 2020 Goa, India, pp. 185–190. Springer Singapore (2021).

Rahman, M., D. Islam, R. J. Mukti, and I. Saha. "A deep learning approach based on convolutional LSTM for detecting diabetes." *Computational Biology and Chemistry* 88 (2020): 107329.

Reza, M. S., R. Amin, R. Yasmin, W. Kulsum, and S. Ruhi. "Improving diabetes disease patients classification using stacking ensemble method with PIMA and local healthcare data." *Heliyon* 10, no. 2 (2024): e24536.

Sánchez-DelaCruz, E., R. Weber, R. R. Biswal, J. Mejia, G. Hernández-Chan, and H. Gómez-Pozos. "Gait biomarkers classification by combining assembled algorithms and deep learning: results of a local study." *Computational and Mathematical Methods in Medicine* 2019, no. 1 (2019): 3515268

Suma, K. V., S. Selvi, P. Nanda, M. Shetty, M. Vikas, and K. Awasthi. "Deep learning approach to nailfold capillaroscopy based diabetes mellitus detection." *International Journal of Online and Biomedical Engineering* (*iJOE*) 18, no. 06 (2022): 95–109.

Swapna, G., K. P. Soman, and R. Vinayakumar. "Automated detection of diabetes using CNN and CNN-LSTM network and heart rate signals." *Procedia Computer Science* 132 (2018): 1253–1262.

Utomo, C. P., M. Fathurahman, and D. F. D. Saputra. "Diabetes prediction of critical care patient using catboost algorithm." *AIP Conference Proceedings* 2838, no. 1. AIP Publishing, Surakarta, Indonesia (2024).

Zanelli, S., M. A. El Yacoubi, M. Hallab, and M. Ammi. "Type 2 diabetes detection with light CNN from single raw PPG wave." *IEEE Access* 11 (2023): 57652–57665.

CHAPTER 10

Deep Learning in Diabetes Mellitus Detection and Diagnosis

Atianashie A. Miracle and
Chukwuma Chinaza Adaobi

10.1 INTRODUCTION

To begin with, the term "diabetes mellitus" is introduced. According to the authors, diabetes mellitus is a group of metabolic disorders characterized by prolonged high blood sugar and disturbances of carbohydrate metabolism. It is crucial to have an accurate prognosis of diabetes mellitus, as it will help physicians prevent the development of important microvascular and macrovascular complications (Rajagopal et al., 2022; Wei et al., 2022). In the modern era of "Evidence-Based Medicine," it would be unusual to undertake a new intervention or treatment without good scientific evidence of its effectiveness. It differs from prognosis, which is the possible outcomes of a disease.

We aim to discover the patterns in the data that are most predictive of the possible risks, complications, and their occurrence in the progress of the disease. In recent years, deep learning has gained increasing attention and popularity in the field of prognostication, particularly in terms of diagnosis, treatment selection, and outcome prediction. Deep learning is a specific type of machine learning in which the computer is made to think

DOI: 10.1201/9781032708430-10

about how the human brain works, i.e., it works on the modular hierarchies of the concept in improving the number of concept layers. It allows the computer to be able to interact with the surroundings and also makes the computer learn from the data that had been presented before (Huang et al., 2015; Rawat et al., 2022; Suyanto et al., 2022). This will enable the computer to recognize the patterns of the diagnosis and help in the early detection of certain conditions. However, traditional machine learning algorithms such as support vector machines or random forests require expert input for the feature selection part. Deep learning allows involving minimum human effort in feature extraction due to its highly hierarchical feature learning (Maniruzzaman et al., 2017). It is evident that deep learning models, unlike classical multivariable regression models, can efficiently use the raw datasets without the need for the prespecification of the variables for analysis.

10.1.1 Definition of Diabetes Mellitus

Diabetes mellitus is a chronic disease that occurs either when the pancreas does not produce enough insulin or when the body cannot effectively use the insulin it produces. Insulin is a hormone that regulates blood sugar. The disease can lead to dangerous complications. There are three main types of diabetes mellitus: type 1, type 2, and gestational. Type 1 diabetes is characterized by deficient insulin production. It is caused by an autoimmune reaction in which the body's defense system attacks the cells that produce insulin. Type 2 diabetes is much more common than type 1 and is often linked to the genes we inherit and our lifestyle choices. The most common symptoms of diabetes, regardless of type, are increased blood sugar levels and frequent urination. As time goes on, the body's insulin resistance can lead to a reduction in the efficiency of the insulin receptors. Insulin resistance is the main factor driving the disease (Li et al., 2022). Over time, the pancreas may lose its ability to produce insulin. This lack of response in the insulin receptors combined with the decrease in insulin production contributes to chronic hyperglycemia.

Diabetes is a major public health crisis. As of 2019, the International Diabetes Federation (IDF) estimated that over 460 million people worldwide were living with diabetes. Moreover, the World Health Organization reports that the condition is a leading cause of death and disability, with an estimated 1.6 million deaths globally directly attributed to diabetes in 2016. Cultivating an understanding of the pathophysiological events associated with diabetes mellitus provides strategic targets for discovering a means of prevention or management. The next step in finding effective treatments, and ultimately, a cure for diabetes is research. Sports scientist Stephen Bradley says, "We know today more about the genes of a worm that lived 30 years ago than we know about human genes today (Marques et al., 2022; Singh & Singh, 2020). There's so much progress to be made." And progress is being made, thanks in part to the diabetes research community. With the help of studies using different techniques and taking different angles, readers are getting to know more about diabetes-related statistical science. And there are many different types of statistics researchers in the diabetes field. Some of them focus on methodologies, such as survival analysis or pattern recognition. Others concentrate more on applied work, analyzing clinical trial data for example. Each of the statistics work plays an important role in improving our understanding of the disease as well as our ability to prevent and treat diabetes. This chapter focuses on the use of deep learning in the prognosis of diabetes mellitus (Mohan et al., 2013; Nuankaew et al., 2021).

10.1.2 Importance of Prognosis in Diabetes Mellitus

The translation of deep learning techniques to diabetes prognosis is a relatively new area. If the promise of this rapidly developing field is to be realized for patients with diabetes and their clinicians, it is important to address some of the specific challenges posed by diabetes and its prognosis. This includes the variability in the type and quality of the data that may be collected, understanding how to incorporate different sources of data, and aligning the use of new technologies with existing practices in healthcare. It is certainly an exciting time for the field and, with the proper collaboration between clinicians, researchers, and data scientists, the authors believe that deep learning has the potential to significantly enhance our ability to predict and prevent the complications of diabetes (Opitz & Maclin, 1999; Podgorelec et al., 2002).

Deep learning models, in particular, are designed to learn from the data provided iteratively. By using increasingly complex algorithms and structures, deep learning models can identify intricate patterns in data that may not be immediately obvious to human investigators. This presents an exciting opportunity for improving the accuracy of prognostic models and, crucially, to move closer to achieving personalized prognosis for patients with diabetes mellitus. Deep learning models have already demonstrated significant improvements in the accuracy of predicting outcomes in several conditions (Gnanadass, 2020; Hasan et al., 2020). By recognizing important signals and patterns in complex and large datasets, these models can identify previously unseen or unknown connections between predictor variables and outcomes.

Although the clinical assessment of prognosis is traditionally informed by a range of demographic, clinical, and laboratory variables, the use of prognostic models in the care of patients with diabetes has been limited. This is partly due to the complexity of available models and a general lack of familiarity with their use in routine practice. However, the increasing availability of big health data, defined as extremely large datasets that may be analyzed computationally to reveal patterns, trends, and associations, has opened up new possibilities in the field of prognostic research (Sivaranjani et al., 2021). In particular, the use of machine learning approaches, such as deep learning, has become an area of active interest. This novel branch of statistical learning research has been heralded as having the potential to revolutionize many fields, including medicine.

The accurate estimation and understanding of prognosis are essential in the management of patients with diabetes. In medicine, the term "prognosis" denotes the prediction of the probable course, duration, and outcome of a disease. For patients with diabetes mellitus, prognosis helps to guide decisions regarding the choice of therapy and to identify the need for referral to specialist care. Importantly, it also provides patients with information about the potential impact of the disease on their quality and length of life. Given the increasing prevalence of diabetes and the growing number of potential complications that patients face, prognosis has never been more important.

10.1.3 Role of Deep Learning in Prognosis

Diabetes mellitus has been seen as a complex and heterogeneous metabolic disorder and its prognosis could involve the prediction of different outcomes, ranging from acute events like hypoglycemia to chronic complications like retinopathy. For all these different outcomes as well as the potential factors that might be associated with disease progression, it would be infeasible for traditional methods to manually test and identify all possible interactions between different prognostic factors. This is where deep learning could potentially play a crucial role because of its ability to identify complex interactions and its flexibility in handling both standard clinical and more advanced data (Naz & Ahuja, 2020).

On the other hand, deep learning models are capable of automating the process of identifying complex interactions between different prognostic factors and have the potential to learn from both standard clinical and more cutting-edge data, such as genetic information. This can lead to more accurate predictions and less human input in the prognosis of different outcomes of diabetes mellitus. It is mentioned that standard clinical data relates to information that our body produces as a result of a normal physiological process. This includes signs, symptoms, and laboratory measurements. On the contrary, genetic data refers to information about the cause of diseases that lie in our DNA (Serda et al., 2013). However, data of different types also have their complexity and this requires the use of big data technologies such as Hadoop and MapReduce for data storage and processing, as well as knowledge of parallel processing algorithms to exploit the capability of these frameworks.

Traditionally, healthcare professionals have used regression models to identify prognostic factors and create risk scores for different outcomes of diabetes mellitus. However, these models rely heavily on clinicians to use various tests to approximate the values of different prognostic factors and reach a final diagnosis. Also, these models usually assume that the effect of a certain prognostic factor is the same for all given values of other factors, which may often not hold in reality given the complexity of diabetes mellitus and its natural history (Khan et al., 2019; Olisah et al., 2022).

Deep learning, a subset of artificial intelligence (AI), has shown substantial promise in the medical field, particularly in prognosis. Deep learning algorithms are based on the concept of neural networks, which are designed to simulate the way the human brain analyzes and processes information. These algorithms learn to perform a task, such as making a prediction, by being trained on large amounts of data. The ability of deep learning models to identify complex patterns in input data makes them well-suited for the multifactorial nature of diabetes mellitus and its prognosis.

10.2 DATA COLLECTION AND PREPROCESSING

Next, an example of creating a new variable for an interaction term is shown. Two predictor variables are utilized to create an interaction term, and all of these variables, including the new interaction variable, are stored in a new data frame. The 'names ()' function is used to assign names to the new data frame accordingly and it shows that the new variable is added successfully.

After that, a dummy variable is created by collapsing the factors in the diabetes type. The original variable is then removed and the new dummy variables are added to the data set. I have used the model and the boomSDE package in R to perform the feature selection heuristically based on lasso and for each multiple regression, respectively. By examining the summary plot, the optimal model is derived and the main and interactive effects are visualized.

During the cleaning and imputation process, the variable "date" is converted from numeric to date type. Data entries with missing dates are identified and then removed using the 'complete. Cases ()' function. Besides that, there are some near-zero variance predictors found in the data set. Thanks to the caret package, it provides a function 'nearZeroVar' which uses the frequency ratio of the most common value to the frequency of the second most common value. Through this, variables with a 95:5 ratio can be identified and removed from the dataset.

Suppose that we have dummy data utilizing 6 months of data from two sources: primary care (PC) and hospital specialist (HS). A table data frame is created to record the patient's ID, gender, age, diabetes type, measurement date, measurement values, source of measurement, and some other features. All the data are in numeric and factor types so that they can be used in the later stage of the data processing and analysis. The initial data frame consists of six main variables which are ID, gender, age, diabetes type, measurement date, and values.

First, I introduce the data source for data processing and its features. Second, I perform a data checking and cleaning process, which includes the removal of unused variables. Last but not least, I demonstrate feature engineering using the example of creating new variables for interaction terms. All the data preprocessing is done in R since it has powerful data manipulation packages such as "dplyr" and "data.table" which ease the process and make the code look clean and systematic.

10.2.1 Selection of Relevant Variables

To train a machine learning model to predict (or estimate) an outcome, practitioners would examine widely accepted diabetic measures in the literature. From the statistics generated from computer-based data mining, life table studies, or medicine, typically, a very large list of diabetic-related measures could be used. However, theoretical and clinical knowledge would be important for selecting only a subset of relevant measures to construct an efficient and effective prediction model. This is because including irrelevant or redundant measures in a model would increase its complexity and the noise level in the measures, thus hurting the predictive performance. On the other hand, excluding a measure that has a strong association with the disease in the final model would introduce bias and decrease the accuracy of prediction. Therefore, in this study, careful considerations were given to the selection of relevant measures used to predict the disease progression of a specific group of diabetic patients. A total of 15 measures were initially identified by the expert system from the data warehouse. These measures were selected from the domains of demography, medication use, and diabetes-specific and co-morbidity indicators (Khan et al., 2019). A set of association rules was derived from the knowledge discovery process, concerning the probability of the co-existence of the diabetes disease codes. For example, a rule would suggest that for elder patients with diabetes and the diagnosis of prostate disease, there is a 40% chance of having a co-existence of impotence. Such information could provide valuable guidelines for medical practitioners in understanding and drafting personalized care plans for individual patients.

10.2.2 Data Cleaning and Imputation

The completed datasets from multiple imputations will be stacked together for further analysis. Imputed data provides greater statistical power and allows for more complete analysis. However, it is also suggested to include a sensitivity analysis section to address how the results might change if different imputation methods are used. By doing so, the robustness of the findings could be further demonstrated.

We used the mice package in R and employed predictive mean matching that takes into account the non-linear relationships between missing and auxiliary variables. The number of multiple imputed datasets is determined based on the percentage of missing information. After imputation, we checked the imputations and the distributions of imputed values for all variables. Also, it is important to include missing indicators in the analysis to make sure that the missing data mechanism is missing at random. Specifically, it is helpful to compare the missingness between the original variable and the missing indicator. By noticing these important technical details, the high quality of the imputation could be better assured (Naz & Ahuja, 2020).

In this study, we used a method called multiple imputation to handle missing data in a principled way. There are different methods for handling missing data, such as complete case analysis (i.e. only cases with no missing value are used in the analysis) and mean imputation (i.e. for a variable with missing value, fill in the mean of that variable). However, these methods have limitations and could lead to biased results. Multiple imputation is a valid and flexible method for handling missing data. The key idea of multiple imputation is to replace each missing value with a set of possible values to reflect the uncertainty of imputation. Then, the analysis will be performed on each of the completed datasets and the results will be combined to account for the uncertainty due to missing data. This method provides unbiased parameter estimates and consistent statistical power.

The diabetes dataset used in this study contains missing values and inconsistencies. To obtain reliable results, it is important to detect and correct errors in the data. First, we identified and removed duplicate entries in the dataset. Then, we checked the type and range of each variable. For example, date variables are converted to date type and we examined if each date is within a logical range. Visualizations like histograms and boxplots help identify potential outliers.

10.2.3 Feature Engineering

Feature engineering transforms raw data into formats more understandable for machine learning models, enhancing their performance and efficiency. While deep learning models can directly handle raw data, traditional algorithms like random forest require meticulously crafted features for optimal functioning. Feature selection, a critical step, involves choosing the most relevant features to improve model accuracy and reduce complexity. Techniques like the filter and wrapper methods help identify these crucial features. Our work introduced a novel longitudinal feature for deep learning models, offering a fresh perspective on handling time-series data in diabetes research, showcasing the blend of domain knowledge and advanced analytics to refine model inputs (Naz & Ahuja, 2020).

10.3 DEEP LEARNING MODELS

Neural network models, especially for diabetes prognosis, outperform traditional statistical models by capturing complex biological patterns without pre-defined assumptions. These models require careful optimization and a diverse training dataset to achieve high prognostic accuracy. Despite challenges like the need for large, varied datasets, neural networks' potential for personalized diabetes care is significant, inviting further exploration into various neural network architectures and optimization techniques to enhance diabetes prognosis (Hasan et al., 2020).

10.3.1 Neural Network Architectures

Neural networks, inspired by the human brain, are adept at pattern recognition across multiple layers, making them suitable for complex prognosis tasks like diabetes mellitus prediction. While the architecture of neural networks can vary, the core process involves input data passing through multiple layers to predict health outcomes. This flexibility allows for the development of efficient models tailored to diabetes research, promising more accurate predictions and opening avenues for innovation in medical prognosis (Podgorelec et al., 2002).

10.3.2 Convolutional Neural Networks (CNNs)

CNNs represent a leap in diabetes prognosis, learning directly from raw data without the need for manual feature extraction. This approach simplifies model testing and can potentially streamline diabetes research by quickly adjusting to new prediction tasks. Although CNNs are less common in diabetes research compared to other areas, their growing application points to a future where they could be key to advancing prognosis methods (Singh & Singh, 2020).

10.3.3 Recurrent Neural Networks (RNNs)

RNNs excel in processing sequential data by considering the information from previous steps, making them ideal for time-sensitive data analysis like diabetes prognosis. Their unique architecture allows them to remember and utilize past data, a critical feature for accurately predicting disease progression over time. However, challenges like short-term memory are being addressed with advancements in network designs, such as long short-term memory (LSTM) networks, further refining their predictive capabilities (García-Ordás et al., 2021).

10.3.4 Long Short-Term Memory (LSTM)

LSTMs are an evolution of RNNs designed to capture long-term dependencies in data, crucial for accurately predicting diabetes progression based on historical patient data. Their sophisticated gating mechanism helps prevent the vanishing gradient problem, ensuring relevant information is retained throughout the analysis. LSTMs have shown promise in various fields, highlighting their potential to enhance diabetes prognosis through deep learning (Li et al., 2022).

10.4 TRAINING AND EVALUATION

Effective machine learning requires splitting data into training and testing sets to evaluate models accurately without overfitting. Optimization techniques like Adam and strategies such as *k*-fold cross-validation are employed to refine model performance. Our study leverages these methods to ensure our deep learning models are both accurate and reliable, emphasizing the importance of rigorous training and validation processes in medical prognosis research (Singh & Singh, 2020).

10.4.1 Splitting the Dataset

Splitting the dataset into training and testing sets is crucial for unbiased model evaluation, helping to mitigate overfitting and ensure models generalize well to new data. Our method maintains the original data distribution, ensuring the models are trained and tested on representative samples, a fundamental step in developing reliable prognostic tools (Marques et al., 2022).

10.4.2 Model Training and Optimization

Our approach to model optimization includes stratified dataset splitting and exhaustive search techniques like Grid Search for hyperparameter tuning, ensuring the deep learning models are finely tuned for diabetes prognosis. By employing cross-validation and exploring various optimization algorithms, we aim to build robust models capable of making accurate predictions (Huang et al., 2015).

10.4.3 Performance Metrics for Prognosis

Evaluating deep learning models involves metrics such as sensitivity, specificity, and accuracy, alongside AUC-ROC and AUC-PR curves, to gauge their prognostic performance comprehensively. These metrics help in understanding the models' ability to predict diabetes progression, informing the choice of threshold values for real-world applications and guiding future research toward models that balance sensitivity and specificity according to clinical needs (Nuankaew et al., 2021).

10.5 DEEP LEARNING APPLICATIONS IN DIABETES MELLITUS PROGNOSIS

Deep learning involves the use of ever-more complex neural network architectures to search for ever-more subtle and useful patterns, and this type of AI has shown great promise in the field of diabetes prognosis. Deep learning algorithms are increasingly being used to help determine the likelihood of certain outcomes with diabetes, including risks for cardiovascular disease, and to help provide insights into optimizing and personalizing treatment. These types of study typically involve the use of electronic health record data, or else data from patients in clinical trials, and often combine many different sources of data, including genotypic, phenotypic, proteomic, and metabolomic information. Because of the complex and multifactoral nature of diabetes mellitus, deep learning's ability to integrate this wide range of predictors makes it a particularly exciting tool for improving prognosis. This is important because diabetes prognosis or the likely course and outcome of the disease is critical.

Knowing the prognosis helps doctors and patients make decisions about which treatments are most appropriate. It's also an essential element of effectively managing the disease and its complications, as being able to predict which patients are most at risk of particular complications could help drive preventative care and customize treatment strategies to individual patients (Marques et al., 2022; Nuankaew et al., 2021). Currently, prognosis in diabetes includes using risk scales and scores to try to determine the likelihood of different difficulties, such as the UKPDS risk engine used to assess cardiovascular risk, or tests to monitor for early signs of complications, such as the eye examinations to look for signs of diabetic retinopathy (Suyanto et al., 2022). However, ultimately these methods can only give information about the general likelihood of patients developing issues; they are not capable of providing personalized information or insights into the precise risks or appropriate treatments of the individual patient. Deep learning offers an exciting opportunity to change this and to use the huge amount of patient data now available in electronic form to revolutionize the way we approach diabetes prognosis. It may be that shortly, experts in diabetes care could use a combination of conventional methods and advanced deep learning outputs to offer each patient a truly personalized risk assessment, with data specifically tailored to that individual's circumstances and the next steps in treatment justified by their predicted risks.

10.5.1 Early Detection of Diabetic Complications

Many diabetic complications can be prevented or delayed through early detection and appropriate treatment. It has been found that the presence of peripheral neuropathy and vasculopathy is the most highly influential toward eventual amputation in diabetic patients. As such, researchers have been looking into using physiological data for the early detection of complications. With the development of wearable technology, such as continuous glucose monitors, it is feasible to collect longitudinal data to monitor the progression of diabetes and its associated complications.

The application of deep learning in this area centers around the processing of time-series data. Traditional methods use algorithms, such as fast Fourier transform, to derive frequency-domain metrics from time-series waveforms. Such algorithms, however, require significant manual preprocessing to derive the correct features from the waveform and are often not capable of capturing the complex dynamics of the physiology (Li et al., 2022; Mohan et al., 2013). With the implementation of deep learning models, the need for customizable, manual feature extraction has been negated as neural networks are capable of learning both the temporal and composite features directly from the raw physiological waveforms. This is exemplified in the work of researchers comparing deep learning networks to traditional machine learning in their capability to classify retinopathy from retinal fundus images. The study found that not only did the deep learning network result in superior classification performance compared to conventional methods, but the sensitivity of the convolutional neural network employed in the deep learning model was also significantly higher than the traditional algorithm.

10.6 CONCLUSION

This chapter has illuminated the path forward in the critical field of diabetes mellitus prognosis through the lens of deep learning. We have traversed from foundational goals to the detailed exploration of how integrating deep learning with diabetes research can spark transformative advancements in patient care. The fusion of deep learning's powerful algorithms with the nuanced understanding of diabetes opens a new frontier for innovation, setting the stage for groundbreaking developments in the management and treatment of this pervasive disease. As we stand on the precipice of this exciting juncture, the potential of deep learning in reshaping diabetes care is already unfolding, marking the beginning of a new era of enhanced prognosis and personalized treatment strategies. This shift not only promises a brighter future for individuals living with diabetes but also highlights the importance of continuous research and collaboration in harnessing the full potential of technology in healthcare. The journey through the realms of deep learning and diabetes prognosis underscores a collective ambition to refine and redefine the standards of medical care, ensuring that the advancements in AI translate into tangible benefits for patients worldwide. As we peer into the horizon, it's clear that deep learning will continue to be a beacon of innovation, driving us toward a future where diabetes management is more precise, personalized, and promising than ever before.

REFERENCES

García-Ordás, M. T., Benavides, C., Benítez-Andrades, J. A., Alaiz-Moretón, H., & García-Rodríguez, I. (2021). Diabetes detection using deep learning techniques with oversampling and feature augmentation. *Computer Methods and Programs in Biomedicine*, 202. https://doi.org/10.1016/j.cmpb.2021.105968.

Gnanadass, I. (2020). Prediction of gestational diabetes by machine learning algorithms. *IEEE Potentials*, *39*(6), 32–37. https://doi.org/10.1109/MPOT.2020.3015190.

Hasan, M. K., Alam, M. A., Das, D., Hossain, E., & Hasan, M. (2020). Diabetes prediction using ensembling of different machine learning classifiers. *IEEE Access*, *8*, 76516–76531. https://doi.org/10.1109/ACCESS.2020.2989857.

Huang, J., Li, Y. F., & Xie, M. (2015). An empirical analysis of data preprocessing for machine learning-based software cost estimation. *Information and Software Technology*, *67*, 108–127. https://doi.org/10.1016/j.infsof.2015.07.004.

Khan, R. M. M., Chua, Z. J. Y., Tan, J. C., Yang, Y., Liao, Z., & Zhao, Y. (2019). From pre-diabetes to diabetes: Diagnosis, treatments and translational research. *Medicina (Lithuania)*, *55*(9). https://doi.org/10.3390/MEDICINA55090546.

Li, B., Zhang, J., Yang, R., & Li, H. (2022). FM-Net: Deep learning network for the fundamental matrix estimation from biplanar radiographs. *Computer Methods and Programs in Biomedicine*, 220. https://doi.org/10.1016/j.cmpb.2022.106782.

Maniruzzaman, M., Kumar, N., Menhazul Abedin, M., Shaykhul Islam, M., Suri, H. S., El-Baz, A. S., & Suri, J. S. (2017). Comparative approaches for classification of diabetes mellitus data: Machine learning paradigm. *Computer Methods and Programs in Biomedicine*, *152*, 23–34. https://doi.org/10.1016/j.cmpb.2017.09.004.

Marques, R., Andrade De Jesus, D., Barbosa-Breda, J., Van Eijgen, J., Stalmans, I., van Walsum, T., Klein, S., Vaz, P. G., & Sánchez Brea, L. (2022). Automatic segmentation of the optic nerve head region in optical coherence tomography: A methodological review. *Computer Methods and Programs in Biomedicine*, *220*. https://doi.org/10.1016/j.cmpb.2022.106801.

Mohan, V., Amutha, A., Ranjani, H., Unnikrishnan, R., Datta, M., Anjana, R. M., Staimez, L., Ali, M. K., & Narayan, K. M. V. (2013). Associations of β-cell function and insulin resistance with youth-onset type 2 diabetes and prediabetes among Asian Indians. *Diabetes Technology and Therapeutics*, *15*(4), 315–322. https://doi.org/10.1089/DIA.2012.0259.

Naz, H., & Ahuja, S. (2020). Deep learning approach for diabetes prediction using PIMA Indian dataset. *Journal of Diabetes and Metabolic Disorders, 19*(1), 391–403. https://doi.org/10.1007/S40200-020-00520-5.

Nuankaew, P., Chaising, S., & Temdee, P. (2021). Average weighted objective distance-based method for type 2 diabetes prediction. *IEEE Access, 9*, 137015–137028. https://doi.org/10.1109/ACCESS.2021.3117269.

Olisah, C. C., Smith, L., & Smith, M. (2022). Diabetes mellitus prediction and diagnosis from a data preprocessing and machine learning perspective. *Computer Methods and Programs in Biomedicine, 220*, 106773. https://doi.org/10.1016/J.CMPB.2022.106773.

Opitz, D., & Maclin, R. (1999). Popular ensemble methods: An empirical study. *Journal of Artificial Intelligence Research, 11*, 169–198. https://doi.org/10.1613/JAIR.614.

Podgorelec, V., Kokol, P., Stiglic, B., & Rozman, I. (2002). Decision trees: An overview and their use in medicine. *Journal of Medical Systems, 26*(5), 445–463. https://doi.org/10.1023/A:1016409317640.

Rajagopal, A., Jha, S., Alagarsamy, R., Quek, S. G., & Selvachandran, G. (2022). A novel hybrid machine learning framework for the prediction of diabetes with context-customized regularization and prediction procedures. *Mathematics and Computers in Simulation, 198*, 388–406. https://doi.org/10.1016/j.matcom.2022.03.003.

Rawat, V., Joshi, S., Gupta, S., Singh, D. P., & Singh, N. (2022). Machine learning algorithms for early diagnosis of diabetes mellitus: A comparative study. *Materials Today: Proceedings, 56*, 502–506. https://doi.org/10.1016/j.matpr.2022.02.172.

Singh, N., & Singh, P. (2020). Stacking-based multi-objective evolutionary ensemble framework for prediction of diabetes mellitus. *Biocybernetics and Biomedical Engineering, 40*(1), 1–22. https://doi.org/10.1016/j.bbe.2019.10.001.

Sivaranjani, S., Ananya, S., Aravinth, J., & Karthika, R. (2021). Diabetes Prediction using Machine Learning Algorithms with Feature Selection and Dimensionality Reduction. *2021 7th International Conference on Advanced Computing and Communication Systems, ICACCS 2021*, 141–146. https://doi.org/10.1109/ICACCS51430.2021.9441935.

Suyanto, S., Meliana, S., Wahyuningrum, T., & Khomsah, S. (2022). A new nearest neighbor-based framework for diabetes detection. *Expert Systems with Applications, 199*. https://doi.org/10.1016/j.eswa.2022.116857.

Wei, H., Sun, J., Shan, W., Xiao, W., Wang, B., Ma, X., Hu, W., Wang, X., & Xia, Y. (2022). Environmental chemical exposure dynamics and machine learning-based prediction of diabetes mellitus. *Science of the Total Environment, 806*. https://doi.org/10.1016/j.scitotenv.2021.150674.

CHAPTER 11

Deep Learning Algorithms for Diabetes Mellitus Detection and Management

Mohadeseh Zarei Ghobadi,
Mohammad Dehghani, and Elaheh Afsaneh

11.1 INTRODUCTION

Diabetes mellitus is a long-term metabolic disorder characterized by elevated levels of glucose in the blood. It is mainly classified into three common types: Type 1 diabetes mellitus (T1DM), Type 2 diabetes mellitus (T2DM), and gestational diabetes mellitus (GDM). Each type is indicated by dysregulation of lipids, protein metabolism, and carbohydrate [1]. Furthermore, there are specific types of diabetes, such as neonatal diabetes, maturity-onset diabetes of the young, and diabetes that occur under the effect of exocrine pancreas disorders or chemical/drug-induced causes [2]. Environmental factors, viral infections, and gestational infections can contribute to the progression of T1DM [3]. T2DM has a heritable component, and the risk of developing T2DM is higher among siblings of affected individuals. Maternal history of T2DM and abnormal fasting glucose concentration or high body mass index (BMI) also increases the

DOI: 10.1201/9781032708430-11

risk [4]. GDM refers to glucose intolerance that is first recognized during pregnancy. It affects approximately 7% of pregnancies and poses risks to both the mother and the fetus [5].

Early detection of diabetes mellitus is crucial for implementing lifestyle modifications and interventions to prevent or delay its onset. Various screening methods, including HbA1c, fasting plasma glucose (FPG), and oral glucose tolerance test (OGTT) can be used for identifying individuals with prediabetes or undiagnosed diabetes [6]. On the other hand, computational methods, including machine and deep learning methods, can be helpful for diagnosis, detection, and other aspects related to this disease.

11.2 DEEP LEARNING

Deep learning, within the context of machine learning, is a distinct and specialized domain that revolves around the creation and training of artificial neural networks (ANNs) consisting of multiple layers. These networks extract abstract features from raw input, enabling accurate predictions and informed decisions. Deep learning models acquire hierarchical representations of data, recognizing complex patterns and relationships efficiently. This capacity leads to profound insights and accurate interpretations [7]. In contrast to traditional machine learning methods that necessitate domain-specific knowledge and extensive feature engineering, deep learning autonomously extracts pertinent features, streamlining the process of model development. Deep learning models can directly learn informative representations from raw data, including images, text, or audio, across different modalities [8].

In this chapter, we survey the utilization of diverse deep learning algorithms in different areas of diabetes, such as diagnosis, prediction, treatment, and monitoring. Convolutional neural networks (CNNs), recurrent neural networks (RNNs), deep belief networks (DBNs), and transfer learning are prevalent deep learning techniques employed in various aspects of diabetes engineering, which we discuss in the subsequent sections.

11.2.1 Convolutional Neural Networks

CNNs are a category of sophisticated machine learning models primarily utilized for analyzing visual data. They aim to autonomously identify and extract meaningful features from images or similar grid-like input data. The architecture and training process of CNNs is pivotal in achieving precise and dependable results when employed for diabetes detection. Below is a comprehensive overview of the CNN architecture and training specifically tailored for diabetes detection [7]:

(i) The input layer of a CNN receives various types of data, such as retinal images or clinical records associated with diabetes; (ii) Convolutional layers employ learnable filters to extract significant features from the input data.; (iii) Activation functions introduce non-linearity to capture complex relationships between features; (iv) Pooling layers are an important layer in CNN systems that decreases the dimensionality of the feature maps; (v) Fully connected layers connect neurons and learn high-level representations for output predictions; (vi) A loss function is selected to quantify the discrepancy between predicted outputs and true labels; (vii) Backpropagation and gradient descent are performed to update weights and minimize the loss function; (viii) Training iterations involve repeatedly feeding data through the network, calculating gradients, and updating weights; (ix) Hyperparameters are adjusted to optimize the model's performance; (x) The trained model is assessed on a separate test dataset employing evaluation metrics like accuracy and precision; (xi) Once satisfactory performance is achieved, the trained CNN can be deployed for diabetes diagnosis in a production environment, ensuring consistent preprocessing of input data.

11.2.1.1 Application of CNNs for Various Aspects of Diabetes Management

Until now, CNNs have been applied to multiple aspects of diabetes research and management. Here are some specific areas where CNNs have shown promise:

- **Diabetic retinopathy detection:** CNNs analyze retinal images to detect and classify diabetic retinopathy stages, enabling early intervention [9,10].
- **Glucose monitoring:** CNNs analyze glucose measurements and time-series data to predict blood glucose levels, facilitating personalized monitoring and insulin dosing [11–13].
- **Risk prediction and management:** CNNs predict the risk of developing diabetes like T2DM or complications using various data sources, aiding in personalized risk assessments and early interventions [14].
- **Diabetic foot ulcer detection:** CNNs detect and classify diabetic foot ulcers from images, enabling early detection and monitoring to mitigate complications [15].

11.2.2 Recurrent Neural Networks

RNNs are a kind of deep learning architecture suitable for handling sequential data, including diabetes time-series data. RNNs can capture temporal dependencies and learn from previous inputs to predict or generate sequences. The steps involved in the architecture and training of RNNs for diabetes detection are as follows [7,16,17]:

(i) Inputs: RNNs take sequential data like glucose measurements or insulin doses. The data is preprocessed and represented as input sequences; **(ii) Sequence padding:** Varying length sequences are padded to ensure equal size, enabling batch processing; **(iii) Hidden state:** The hidden state captures the network's memory and previous sequence information, modeling sequential dependencies; **(iv) Recurrence:** At each time step, RNNs update the hidden state by incorporating the input data and the previous hidden state; **(v) Training:** RNNs are trained using a suitable loss function, comparing predicted output with ground truth labels. Weights and biases are adjusted through backpropagation and gradient descent; **(vi) Backpropagation through time (BPTT):** BPTT extends backpropagation to RNNs, propagating errors through all time steps. The remaining steps in the training process are similar to CNNs.

11.2.2.1 Application of RNNs for Various Aspects of Diabetes Management

The performance of RNNs in diabetes-related tasks is influenced by various factors such as the quality and quantity of data, model architecture, feature engineering, and optimization techniques. However, RNNs have demonstrated their potential in improving diabetes management and aiding decision-making processes. Here is how RNNs work for diabetes-related tasks:

- **Glucose level prediction:** RNNs use past glucose readings to predict future levels, capturing patterns and dynamics [18,19].
- **Anomaly detection:** RNNs detect abnormal glucose patterns by learning standard behavior, aiding in identifying episodes or unusual patterns [20,21].
- **Diabetes risk assessment:** RNNs analyze longitudinal data to predict diabetes risk and complications by identifying patterns and risk factors [22].

- **Disease progression modeling:** RNNs analyze longitudinal data to model disease progression, considering demographic information, clinical measurements, and treatment history [23].

11.2.3 Deep Belief Networks

DBNs are valuable algorithms for diabetes management, integrating diverse data modalities like clinical, genetic, and lifestyle factors. By employing unsupervised learning, DBNs capture the relationships and structure within the data, facilitating risk assessment and disease prediction.

The key steps involved in DBNs are as follows [24,25]: **(i) Inputs**: Numerical features related to individuals, including age, gender, BMI, blood pressure, glucose level, etc., are selected and preprocessed; **(ii) Visible layers:** Restricted Boltzmann machines (RBMs) with binary units represent observed variables and connect to the input data; **(iii) Hidden layers:** RBMs, except the last one, have hidden layers that capture latent representations; **(iv) RBM architecture:** Weighted connections exist between visible and hidden layers, with activation functions determining hidden unit activations; **(v) Training:** DBNs undergo unsupervised training, initially employing layer-wise pretraining, followed by fine-tuning using supervised learning algorithms to adjust the biases and weights of the network; **(vi) Inference:** Trained DBNs are used for diabetes detection. Input data passes through visible and hidden layers, producing output used for classification or prediction tasks.

11.2.3.1 Application of DBNs for Various Aspects of Diabetes Management

DBNs have been explored in multiple aspects of diabetes detection, including risk prediction, early detection, classification, and feature selection. Here are some key areas where DBNs have been applied:

- **Early detection:** DBNs analyze longitudinal health records and biomarker changes to detect early stages of diabetes by capturing subtle patterns and trends [26,27].

- **Diabetes management:** DBNs identify critical indicators for diabetes diagnosis, such as glucose concentration, BMI, and age. They provide recommendations for prevention education, personalized measures, and community-wide strategies to improve management [28].

- **Diabetes prediction:** DBN-based models predict the risk of developing diabetes using clinical features like age, BMI, blood pressure, and glucose levels [29].

11.2.4 Transfer Learning

Transfer learning is a valuable technique that leverages pre-trained deep learning models to carry out diagnosis and risk assessment tasks in diabetes. It boosts the performance of models that have limited data by utilizing the knowledge gained from pre-trained models. The architecture of this method in diabetes can be summarized as follows [30,31]: **(i) Source task:** A labeled dataset (S) for the source task, containing input features (x) and target labels (y) indicating diabetes presence or absence; **(ii) Pre-trained model:** A pre-trained model (M) trained on the source task, represented by a function (f_M) with parameters (θ_M); **(iii) Target task:** A labeled dataset (T) for the target task, with input features (x) and target labels (y); **(iv) Fine-tuning:** The pre-trained model (M) is fine-tuned on the target task (T) by updating its parameters (θ_M) to adapt to the new task. The fine-tuned model is denoted as (f_T) with updated parameters (θ_T); **(v) Feature extraction:** Input features (x) from the source task dataset (S) are passed through the pre-trained model (M) to extract high-level features (h); **(vi) Additional layers:** Extra layers are added to process the extracted features (h) and learn task-specific representations. The output is denoted as (h') with parameters (θ'); **(vii) Classification layer:** A final layer predicts the target label (y) for the target task, mapping the output of the additional layers (h') to the target label. The output is denoted as (y') with parameters (θ''); **(viii) Fine-tuning process:** Parameters ($\theta_M, \theta', \theta''$) are updated based on the labeled dataset of the target task (T) using optimization algorithms like stochastic gradient descent (SGD); **(ix) Evaluation and testing:** The fine-tuned model is evaluated on a validation set, and hyperparameters are adjusted as needed. Finally, the model is tested on an independent dataset to assess its performance on unseen data from the target task.

11.2.4.1 Application of Transfer Learning for Various Aspects of Diabetes Management

- **Blood glucose prediction:** It predicts patient-specific blood glucose levels using continuous glucose monitoring data, enabling personalized glucose management and control [32].

- **Diabetic retinopathy detection:** Transfer learning improves the accuracy of automated diagnosis and classification of diabetic retinopathy from retinal images, resulting in enhanced efficiency in identifying retinal abnormalities [33,34].
- **Risk stratification:** Transfer learning assesses the risk of complications in individuals with diabetes by integrating various patient data, enabling early intervention and targeted management strategies to improve outcomes [35].

11.3 FUTURE SCOPES

Applying various deep learning methods in diabetes management offers several promising future perspectives. Here are some key areas where deep learning can play a significant role:

1. **Personalized glucose control:** Models provide tailored recommendations for insulin dosing, meal planning, and exercise routines, optimizing glucose control.
2. **Decision support systems:** Deep learning integrated into systems offers evidence-based recommendations and alerts for better treatment decisions.
3. **Risk stratification and complication prediction:** Deep learning identifies high-risk individuals, enabling early interventions and targeted preventive measures.
4. **Behavioral interventions and adherence monitoring:** Deep learning monitors behavior, and adherence, and provides personalized feedback for effective management.
5. **Telemedicine and remote monitoring:** Deep learning algorithms analyze wearable device data, improving access to care and patient outcomes.
6. **Data integration and knowledge discovery:** Deep learning analyzes diverse datasets, discovering new associations and targets for personalized treatment.
7. **Patient education and empowerment:** Deep learning develops intelligent tools for personalized education, empowering patients to make informed decisions.

These perspectives highlight deep learning's potential in diabetes management, personalized care, prevention, and patient empowerment. Advancements in techniques and data availability will enhance their role in diabetes care.

11.4 DISCUSSION

Deep learning, a subset of artificial intelligence, shows great promise in diabetes detection and diagnosis by analyzing diverse datasets. By analyzing complex patterns and relationships within data, including factors such as age, BMI, family history, and biomarkers, deep learning techniques, especially CNNs, can predict an individual's risk of developing diabetes. Additionally, they excel in automating the detection of diabetic retinopathy from retinal images, facilitating timely intervention. For optimal glycemic control, deep learning models analyze insulin dosages, continuous glucose monitoring data, and dietary information to forecast future blood glucose levels. This assists in making informed decisions about insulin dosing, dietary adjustments, and lifestyle modifications. Deep learning algorithms also aid in the early detection of complications like neuropathy and nephropathy by analyzing various data sources, enabling timely interventions and preventive measures.

Furthermore, deep learning models generate personalized treatment recommendations by considering individual patient data, including lifestyle factors, medical history, genetic information, and treatment response. This personalized approach has the potential to improve treatment outcomes and patient satisfaction.

Despite the progress made, there are still hurdles to overcome when applying deep learning to diabetes. These challenges include the requirement for large, diverse, and annotated datasets, ensuring the reliability and interpretability of models, and addressing algorithm bias and data privacy concerns. In summary, deep learning offers solutions to challenges in diabetes detection and diagnosis. It aids in risk prediction, retinopathy detection, glycemic control, complication detection, and personalized treatment. Overcoming challenges will further enhance its impact on diabetes management.

11.5 CONCLUSION

Deep learning techniques offer significant potential in detecting and diagnosing diabetes mellitus. They can predict diabetes risk, automate retinopathy screening, and forecast glucose levels, aiding healthcare professionals in delivering personalized care and early intervention. However, further research, collaborations, and integration into clinical practice are needed to leverage these technologies fully. Overcoming challenges and advancing the field will lead to improved diabetes management and better patient outcomes in the future.

REFERENCES

1. DeFronzo RA, Ferrannini E, Groop L, Henry RR, Herman WH, Holst JJ, et al. Type 2 diabetes mellitus. *Nature Reviews Disease Primers* 2015;1(1):1–22.
2. Association AD. 2. Classification and diagnosis of diabetes: standards of medical care in diabetes—2021. *Diabetes Care* 2021;44(Supplement_1):S15–S33.
3. Beyerlein A, Donnachie E, Jergens S, Ziegler A-G. Infections in early life and development of type 1 diabetes. *JAMA* 2016;315(17):1899–901.
4. Lyssenko V, Almgren P, Anevski D, Perfekt R, Lahti K, Nissén M, et al. Predictors of and longitudinal changes in insulin sensitivity and secretion preceding onset of type 2 diabetes. *Diabetes Care* 2005;54(1):166–74.
5. Choudhury AA, Rajeswari VDJB. Gestational diabetes mellitus-a metabolic and reproductive disorder. *Biomedicine & Pharmacotheraphy* 2021;143:112183.
6. Afsaneh E, Sharifdini A, Ghazzaghi H, Ghobadi MZJD, Syndrome M. Recent applications of machine learning and deep learning models in the prediction, diagnosis, and management of diabetes: a comprehensive review. *Diabetology & Metabolic Syndrome* 2022;14(1):1–39.
7. Goodfellow I, Bengio Y, Courville A. *Deep Learning*, Cambridge, MA: MIT Press, 2016.
8. Hatcher WG, Yu W. A survey of deep learning: platforms, applications and emerging research trends. *IEEE Access* 2018;6:24411–32.
9. Ting DSW, Cheung CY- L, Lim G, Tan GSW, Quang ND, Gan A, et al. Development and validation of a deep learning system for diabetic retinopathy and related eye diseases using retinal images from multiethnic populations with diabetes. *JAMA* 2017;318(22):2211–23.
10. Chia MA, Hersch F, Sayres R, Bavishi P, Tiwari R, Keane PA, et al. Validation of a deep learning system for the detection of diabetic retinopathy in Indigenous Australians. *British Journal of Ophthalmology* 2024;108(2):268–73.
11. Jaloli M, Cescon M. Long-term prediction of blood glucose levels in type 1 diabetes using a CNN-LSTM-based deep neural network. *Journal of Diabetes Science Technology* 2023;17(6):1590–601.
12. Alvarado J, Velasco JM, Chavez F, Fernández-de-Vega F, Hidalgo JI. Combining wavelet transform with convolutional neural networks for hypoglycemia events prediction from CGM data. *Chemometrics Intelligent Laboratory Systems* 2023;243:105017.

13. Lee JE, Jeon H-J, Lee O-J, Lim HG. Diagnosis of diabetes mellitus using high frequency ultrasound and convolutional neural network. *Ultrasonics* 2024;136:107167.
14. Alqushaibi A, Hasan MH, Abdulkadir SJ, Muneer A, Gamal M, Al-Tashi Q, et al. Type 2 diabetes risk prediction using deep convolutional neural network based-Bayesian optimization. *Computers, Materials Continua* 2023;75(2): 3223–3238.
15. Alzubaidi L, Fadhel MA, Oleiwi SR, Al-Shamma O, Zhang J. DFU_QUTNet: diabetic foot ulcer classification using novel deep convolutional neural network. *Multimedia Tools Applications* 2020;79(21–22):15655–77.
16. Sherstinsky AJ. Fundamentals of recurrent neural network (RNN) and long short-term memory (LSTM) network. *Physica D: Nonlinear Phenomenaa* 2020;404:132306.
17. Dhruv P, Naskar S. Image classification using convolutional neural network (CNN) and recurrent neural network (RNN): a review. *Machine Learning Information Processing: Proceedings of ICMLIP* 2020:367–81.
18. Zhu T, Li K, Chen J, Herrero P, Georgiou P. Dilated recurrent neural networks for glucose forecasting in type 1 diabetes. *Journal of Healthcare Informatics Research* 2020;4:308–24.
19. Martinsson J, Schliep A, Eliasson B, Mogren O. Blood glucose prediction with variance estimation using recurrent neural networks. *Journal of Healthcare Informatics Research* 2020;4:1–18.
20. Doike T, Hayashi K, Arata S, Mohammad KN, Kobayashi A, Niitsu K, editors. A blood glucose level prediction system using machine learning based on recurrent neural network for hypoglycemia prevention. *2018 16th IEEE International New Circuits and Systems Conference (NEWCAS)*, 2018, IEEE, Montreal, QC, Canada.
21. Kim D-Y, Choi D-S, Kim J, Chun SW, Gil H-W, Cho N-J, et al. Developing an individual glucose prediction model using recurrent neural network. *Sensors* 2020;20(22):6460.
22. Cui S, Li C, Chen Z, Wang J, Yuan J. Research on risk prediction of dyslipidemia in steel workers based on recurrent neural network and LSTM neural network. *IEEE Access* 2020;8:34153–61.
23. Alaa AM, van der Schaar M. Attentive state-space modeling of disease progression. advances in neural information processing systems. *32: Annual Conference on Neural Information Processing Systems 2019, NeurIPS 2019*, Vancouver, BC, 8–14 December 2019.
24. Hinton GE, Osindero S, Teh Y-W. A fast learning algorithm for deep belief nets. *Neural Computation* 2006;18(7):1527–54.
25. Salakhutdinov R, Hinton G, editors. Deep Boltzmann machines. *Proceedings of the 12th International Conference on Artificial Intelligence and Statistics (AISTATS) 2009*, Clearwater Beach, FL, Volume 5 of JMLR: W&CP 5, 2009, PMLR.
26. Prabhu P, Selvabharathi S, editors. Deep belief neural network model for prediction of diabetes mellitus. *2019 3rd International Conference on Imaging, Signal Processing and Communication (ICISPC)*, 2019, IEEE, Singapore.

27. Rajavel R, Sundaramoorthy B, GRK, Ravichandran SK, Leelasankar K. Cloud-enabled diabetic retinopathy prediction system using optimized deep belief network classifier. *Journal of Ambient Intelligence Humanized Computing* 2023;14(10):14101–9.
28. Zhao Z, Liu Y, Sun H, Han X, Wang R, editors. Research on diabetes aided diagnosis model based on deep belief network. *Wireless and Satellite Systems: 10th EAI International Conference, WiSATS 2019*, Harbin, China, January 12–13, 2019, Springer.
29. Lang L-Y, Gao Z, Wang X-G, Zhao H, Zhang Y-P, Sun S-J, et al. Diabetes prediction model based on deep belief network. *Journal of Computational Methods in Sciences Engineering* 2021;21(4):817–28.
30. Tan C, Sun F, Kong T, Zhang W, Yang C, Liu C, editors. A survey on deep transfer learning. *Artificial Neural Networks and Machine Learning–ICANN 2018: 27th International Conference on Artificial Neural Networks*, Rhodes, Greece, October 4–7, 2018, Proceedings, Part III 27, Springer.
31. Pan SJ, Yang Q. A survey on transfer learning. *IEEE Transactions on Knowledge Data Engineering* 2009;22(10):1345–59.
32. Deng Y, Lu L, Aponte L, Angelidi AM, Novak V, Karniadakis GE, et al. Deep transfer learning and data augmentation improve glucose levels prediction in type 2 diabetes patients. *NPJ Digital Medicine* 2021;4(1):109.
33. Al-Smadi M, Hammad M, Baker QB, Sa'ad A. A transfer learning with deep neural network approach for diabetic retinopathy classification. *International Journal of Electrical Computer Engineering* 2021;11(4):3492.
34. Khalifa NEM, Loey M, Taha MHN, Mohamed HNET. Deep transfer learning models for medical diabetic retinopathy detection. *Acta Informatica Medica* 2019;27(5):327.
35. Lakshmi Prabha P, Jayanthy AK, Prem Kumar C, Ramraj B. Prediction of cardiovascular risk by measuring carotid intima media thickness from an ultrasound image for type II diabetic mellitus subjects using machine learning and transfer learning techniques. *The Journal of Supercomputing* 2021;77: 10289–10306.

CHAPTER 12

An Analysis of Deep Learning Models for Diabetic Retinopathy Detection and Classification Based on Fundus Image

Govind Narayan Patel, Lipika Dinda, Jitesh Pradhan, and B. Ramachandra Reddy

12.1 INTRODUCTION

Millions of people are suffering from diabetes mellitus, a major public health issue that is expected to hit 700 million people by the year 2045 [1]. Diabetic retinopathy (DR) is a very common eye disease associated with diabetes, affecting at least one-third of those with the condition [2]. Regardless of how bad their diabetes is, any patient might develop diabetic retinal disease, which is characterized by gradually developing retinal vascular abnormalities brought on by ongoing hyperglycemia [3]. It is the main reason for blindness among adults in their working years worldwide,

DOI: 10.1201/9781032708430-12

and the estimated number of individuals with DR is 93 million [4]. It is anticipated that these figures will increase even further, mostly due to the growing incidence of diabetes in developing Asian nations (India, China, etc.) [5,6].

Most DR patients are symptomless in the preliminary stages: clinically undetected microvascular abnormalities and neuronal retinal damage are happening at this time [7]. Therefore, diabetic individuals must undergo routine eye screenings since timely diagnosis and medical intervention are essential [8]. Since treating hyperlipidemia, hyperglycemia, and hypertension is the only precautionary measure, it becomes even more important to diagnose DR early [7]. Furthermore, in up to 98% of instances, currently available therapies like diabetic maculopathy and proliferative retinopathy, which are treated in their initial phases, laser-photocoagulation considerably reduces the risk of blindness. Receiving the right care and early detection are essential to delaying or even preventing blindness from DR [9,10]. Although the first-stage diagnosis of DR might be founded on anomalies in function in electro-retinography, retinal blood vessels, and retinal blood flow diameter, in medical practice, the study of fundus images is the main focus of the preliminary diagnostic [11,12]. Fundus images, one of the most often used methods for assessing the degree of DR, are a simple, rapid, accessible, and well-tolerated imaging technique [13]. Fundus images allow ophthalmologists to see retina lesions in high resolution, which they use to confirm and assess DR disease. However, hand-diagnosing DR from the fundus dataset images necessitates a high degree of experience and output from an experienced eye doctor, especially in remote or densely inhabited regions such as India and Africa, where the overall number of people with DR and diabetes is anticipated to rise sharply in the coming years and ophthalmologists are disproportionately few [14–17]. This has sparked the development of computer-aided diagnosis technologies, which will cut down on the amount of money, work, and time that a medical professional must spend to identify DR. It is now possible to design deep learning systems for precise recognition of DR and categorization because of recent advances in artificial intelligence (AI) and the expansion of computational capabilities and resources. Recent deep learning-based techniques, that is, those released after 2016, for the categorization and detection of DR are described and evaluated here.

In the past few years, a few review articles have been published on the application of deep learning techniques to deep reinforcement learning [18–22], but most of them only address particular facts of the modeling

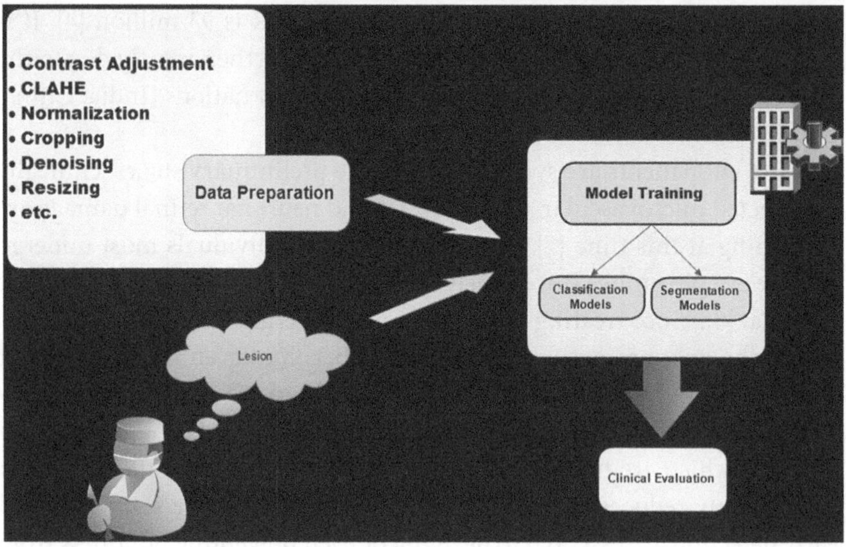

FIGURE 12.1 Analysis of image dataset.

pipeline and the data analysis, as shown in Figure 12.1. This includes, for example, reporting model performance [21,22] or common pre-processing techniques [19,20], while Ref. [22] omits a thorough description of the publicly available assets. A more thorough and coordinated effort is needed to assess the technological implementations and advancements in this extremely busy field of study, as these dispersed activities need. To achieve this, a comprehensive overview of the pipeline for analysis has been presented in Figure 12.1, which includes a detailed analysis of the datasets and the commonly utilized preparatory processing pipelines, also presenting models that have been utilized in real-world medical contexts and offering comparative technical details about the formation of published deep learning models for fundus image classification and segmentation.

12.2 DIABETIC RETINOPATHY

When diabetic retinopathy (DR) first manifests [23], micro-aneurysms are visible on the retina and brought on by pericyte loss and degeneration, which dilate the capillary walls [8]. Intraretinal hemorrhages happen when a microaneurysm or capillary wall ruptures. Hard and soft exudates, venous beading, intraretinal microvascular abnormalities (IRMA), reduplication, or venous loops are further lesions of non-proliferative DR. In regions of ischemia, IRMAs show large caliber tortuous arteries and may be an effort at vascular remodeling, according to Stitt et al. [24]. Lastly, the presence of neo-vascularization – basically the development of new retina vessels from existent ones because of ischemia – is the basis for differentiating between proliferative and non-proliferative DR. Diabetic macular edema (DME) is the most prevalent cause of blindness and can occur in any phase of DR. Edema is linked to abnormalities such as micro-aneurysms or hemorrhages within the fovea center's single disk diameter, exudates that are located within the fovea center's disk diameter, exudates inside the macula, and retinal thickness within the fovea center's disk diameter.

The Early Treatment Diabetic Retinopathy Study, i.e., ETDRS grading system, is regarded as the benchmark for clinical grading processes in DR; however, its practicality and ease of usage in daily clinical practice have not been demonstrated. Hence, to enhance patient screening and caregiver communication, several substitute scales have been put forth. Thus far, there isn't a single worldwide severity scale for DR due to the creation of such streamlined scales in several nations. To do this, the International Clinical Diabetic Retinopathy Disease Severity Scale (ICDRDSS), which divides DR into five levels as indicated in Table 12.1, was suggested by the global diabetic retinopathy project group (Figure 12.2).

TABLE 12.1 Levels in ICDRDSS [23]

S.No.	Severity Level of Disease	Dilated Ophthalmoscopy Findings
1	0 Indicates no diabetic retinopathy	Zero abnormalities
2	1 Indicates mild diabetic retinopathy	Only micro-aneurysms
3	2 Indicates moderate diabetic retinopathy	Greater than micro-aneurysms, but not as much as NPDR (non-proliferative diabetic retina)
4	3 Indicates severe diabetic retinopathy	Any of the subsequent conditions without proliferative retinopathy symptoms: 1. In each of the four quadrants, more than 20 intraretinal hemorrhages. 2. Distinctive venous beading in two or more quadrants. 3. Intraretinal microvascular abnormalities (IRMA) stand out in one or more than one quadrant.
5	4 Indicates growth-oriented diabetic retinopathy	Both or one of the subsequent items: 1. Neo-vascularization 2. Pre-retinal or vitreous hemorrhage

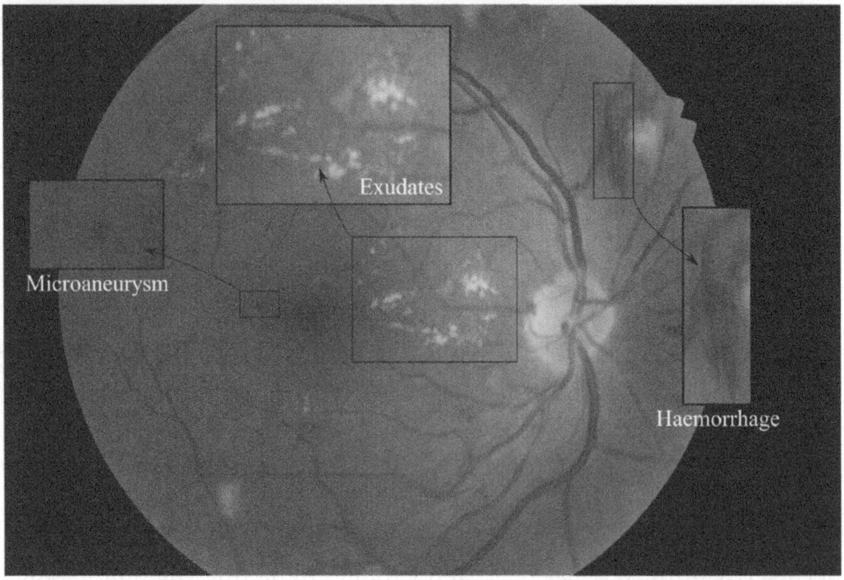

FIGURE 12.2 Indicating diabetic retinopathy on fundus image [23].

12.3 RELATED WORK

Several studies on the subject of automatically detecting DR have been published in various literature. Based on the methods employed, automatic DR detection can be roughly divided into two groups. The first person to use deep learning-based DR detection relies on automated feature extraction, as opposed to typical machine learning algorithms that require the use of methods for external feature extraction using image processing (Table 12.2).

12.4 DATASET DESCRIPTION

This dataset [34] contains 757 fundus images and was obtained at the Department of Ophthalmology in Paraguay (Table 12.3).

12.5 DEEP LEARNING ANALYSIS

G.E. Hinton initially proposed [35] deep learning in 2006 as an integral part of the machine learning process, specifically focusing on DNNs (Deep Neural Networks). These networks, inspired by the human brain and composed of multiple neurons, form an impressive and intricate network. Both supervised and unsupervised categories can be trained by deep learning networks.

TABLE 12.2 Literature Survey of Papers

S. No.	Author	Technique	Dataset	Description
1	Varun Gulshan et al. [25]	Identification of referable diabetics (DR) using deep learning.	Messidor-2 and EyePACS-1 datasets	The algorithm's area under the ROC (Receiver operating characteristic) curve for detecting referable diabetic retinopathy (RDR) was 0.990 (95% CI, 0.986–0.995) for Messidor-2 and 0.991 (0.988–0.993, 95% CI) for EyePACS-1.
2	Kele Xu et al. [26]	Five different transformation types: rotation, flipping, shearing, rescaling, and translation	Color fundus image of the Kaggle dataset	With a 94.5% accuracy rate, they examined the application of the deep CNN approach for diabetic retinopathy.
3	Sheikh Muhammad et al. [27]	A new deep CNN with two fully connected layers and one output layer.	Kaggle EyePACS dataset	The design produced 0.844 AUC and 0.851 QWK scores. It demonstrated 98% sensitivity and 94% specificity in the preliminary stage recognition.
4	Carson Lam et al. [28]	Present the application of CNNs to diabetic retinopathy recognition. Develop a model of multinomial classification.	Color fundus images from the Messidor-1 dataset and Kaggle dataset	The peak test set accuracy for two-ary, three-ary, and four-ary classification models was raised to 74.5%, 68.8%, and 57.2%, respectively, by transfer learning on models AlexNet and GoogLeNet from ImageNet.

(*Continued*)

TABLE 12.2 (*Continued*) Literature Survey of Papers

S. No.	Author	Technique	Dataset	Description
5	Sara Hosseinzadeh Kassani et al. [29]	Four deep CNN extractors (ResNet50, Original Xception architecture, InceptionV3, and MobileNet).	The Kaggle dataset (APTOS 2019)	Improved Xception deep CNN improves DR categorization with a classify-cation accuracy of 83.09% over 79.59%, specificity of 87.00% vs 86.32%, and sensitivity of 88.24% versus 82.35% compared to the original Xception architecture.
6	Anuj Jain et al. [30]	Convolutional neural networks (VGG19, Inception v3 and VGG16)	The EyePACS dataset was made accessible as part of the Kaggle competition.	The accuracy for VGG19 was 76.9%, while that of VGG16 was 71.7%, and that of Inception v3 was 70.2%.
7	BorysTymchenko et al. [31]	Ensemble of three CNN architectures (SE- ResNeXt50, EfficientNet-B4, and EfficientNet-B5)	APTOS 2019 dataset, Kaggle dataset	The proposed approach, which has a quadratic weighted Kappa score of 0.925466, ranks 54 out of 2943 competing approaches with a sensitivity and specificity of 0.99.
8	Akhilesh Kumar Gangwar et al. [32]	A hybrid model has been made using Inception-ResNet-v2 with a customized CNN.	Messidor-1 and APTOS (2019) (Kaggle)	APTOS datasets accuracy is 82.18% and Messidor-1 datasets accuracy is 72.33%
9	Pratt et al. [33]	Employed data augmentation approaches in addition to utilizing 13 CNN layers.	EyePACS DR dataset	Using 80,000 photos of training data, 95% sensitivity was acquired on 5000 images
10	Lam et al. [36]	Transfer learning with models from the ImageNet dataset, including GoogleNet and AlexNet.	Messidor-1 DR dataset	The highest accuracy that GoogleNet has attained is 66.03%

TABLE 12.3 Dataset Description

S.No.	Class Name	No. of Images
1	No diabetic retinopathy signs	187
2	Mild non-PDR (NPDR)	4
3	Moderate non-PDR	80
4	Severe non-PDR	176
5	Very severe non-PDR	108
6	PDR (proliferative diabetic retinopathy)	88
7	Advance proliferative diabetic retinopathy	114

The realm of deep learning encompasses various network types, including RNN (recurrent neural networks), CNN (convolutional neural networks), DBN (deep belief networks), recursive neural networks, and many more. Neural networks play a crucial role in diverse applications such as vector representation, text generation, sentence classification and modeling, word estimation, and feature analysis.

12.5.1 VGG-16

VGG-16 [36] stands out as a CNN architecture meticulously crafted for image classification tasks. Originating from the Visual Geometry Group at Oxford University, the term "VGG-16" specifically denotes a neural network characterized by 16 weight layers, encompassing three fully connected layers and 13 convolutional layers. This architecture gained widespread recognition due to its simplicity and memorable efficacy in image recognition processes. The design principles of VGG-16 include the use of 3×3 convolutional filters with one stride, activation functions employing rectified linear units (ReLU), and the incorporation of 2×2 max-pooling layers with two strides. Renowned for its straightforward yet powerful structure, VGG-16 serves as a benchmark model in the field of deep learning and often serves as a standard for analyzing the performance of other sophisticated neural network models (Figures 12.3, 12.4, and Table 12.4).

12.5.2 VGG-19

The VGG-19 is a CNN architecture. It belongs to the VGG family, which is noted for its simplicity and efficacy in picture classification tasks. VGG-19 is made up of 19 layers, the first 16 of which are convolutional, followed by three fully linked layers. The convolutional layers are divided into five blocks, with every block consisting of numerous convolutional layers followed by max-pooling layers. The implementation of modest 3×3

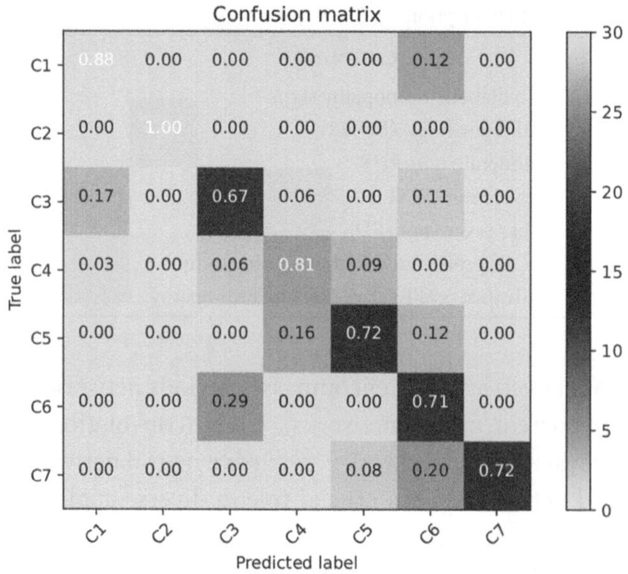

FIGURE 12.3 Confusion matrix of VGG-16.

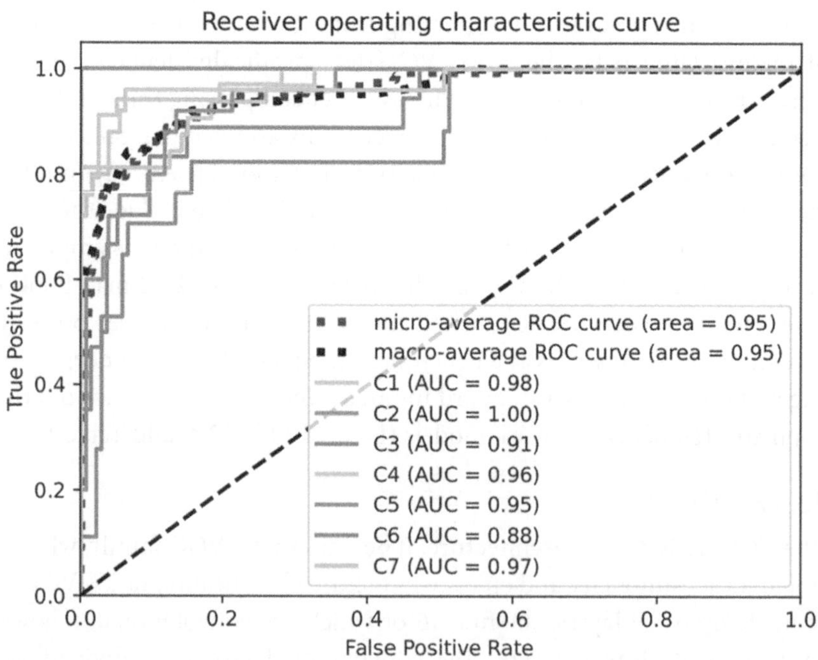

FIGURE 12.4 Receiver operating characteristic curve of VGG-16.

TABLE 12.4 Classification Report of VGG-16

Classes	Precision	Recall	F1-Score	Support
0	0.88	0.88	0.88	34
1	1.00	1.00	1.00	1
2	0.63	0.67	0.65	18
3	0.84	0.81	0.83	32
4	0.78	0.72	0.75	25
5	0.46	0.71	0.56	17
6	1.00	0.72	0.84	25
Micro-average	0.77	0.77	0.77	152
Macro-average	0.80	0.79	0.79	152
Weight-average	0.80	0.77	0.78	152
Samples-average	0.77	0.77	0.77	152

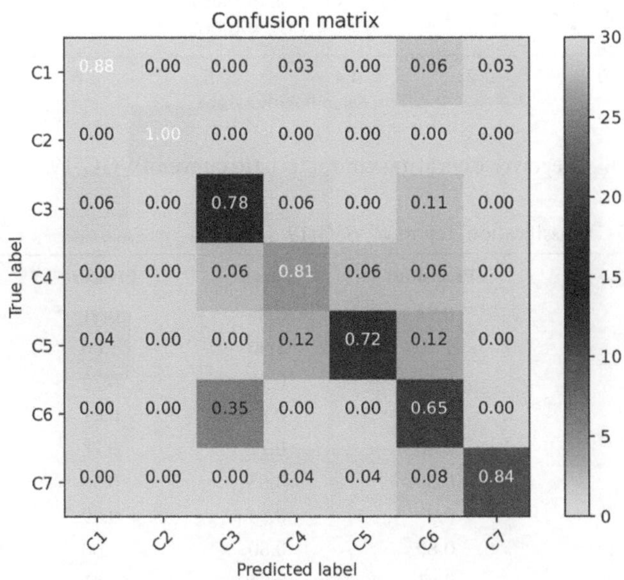

FIGURE 12.5 Confusion matrix of VGG-19.

convolutional filters throughout the network allows for a more complex design while remaining relatively simple. VGG-19 outperformed numerous benchmark datasets, making it a well-studied and cited model in the field of computer vision (Figures 12.5, 12.6, and Table 12.5).

12.5.3 ResNet-50

ResNet-50, a variant of the ResNet [37] architecture, is a CNN distinguished by its 50 layers. This model comprises 48 CNN layers, complemented by

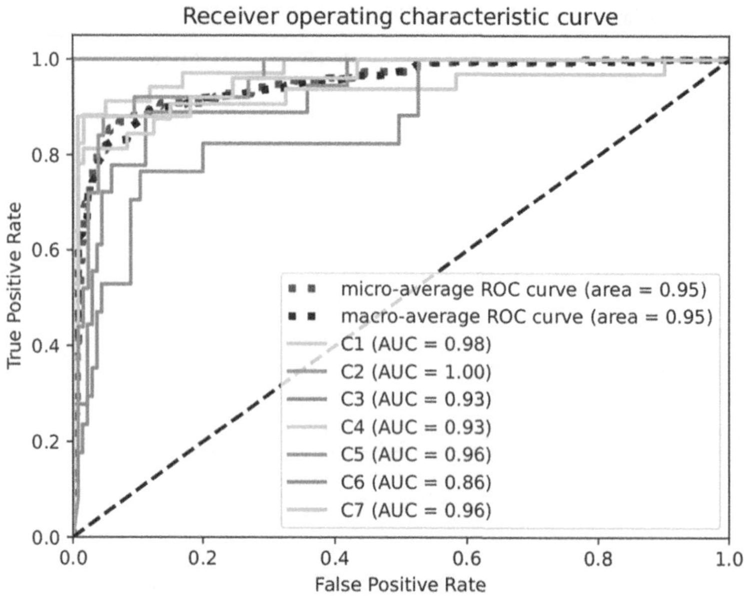

FIGURE 12.6 Receiver operating characteristic curve of VGG-19.

TABLE 12.5 Classification Report of VGG-19

Classes	Precision	Recall	F1-Score	Support
0	0.94	0.88	0.91	34
1	1.00	1.00	1.00	1
2	0.64	0.78	0.70	18
3	0.81	0.81	0.81	32
4	0.86	0.72	0.78	25
5	0.50	0.65	0.56	17
6	0.95	0.84	0.89	25
Micro-average	0.80	0.80	0.80	152
Macro-average	0.81	0.81	0.81	152
Weight-average	0.82	0.80	0.80	152
Samples-average	0.80	0.80	0.80	152

one max-pooling layer and one average pooling layer. Based on the framework for deep residual learning, ResNet-50 effectively addresses the vanishing gradient problem encountered in extremely DNNs. ResNet-50 has slightly more than 23 million training parameters, which is a substantially reduced number compared to other architectures despite its significant depth. The exceptional performance of ResNet-50 is a subject of ongoing discussion, but a fundamental understanding can be gained by exploring the concept of residual blocks and their operational mechanisms (Figures 12.7, 12.8, and Table 12.6).

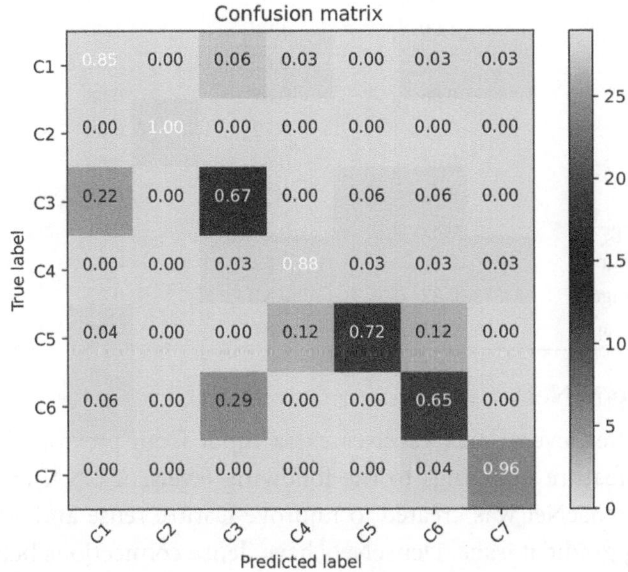

FIGURE 12.7 Confusion matrix of ResNet-50.

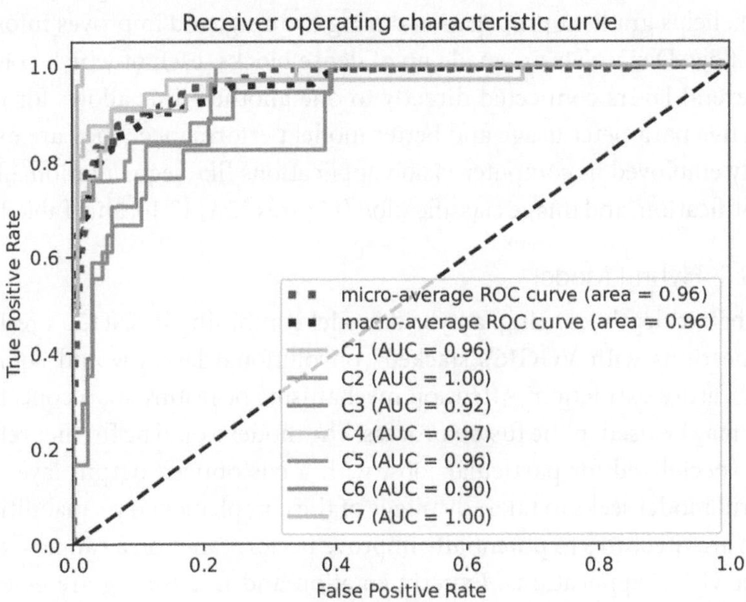

FIGURE 12.8 Receiver operating characteristic curve of ResNet-50.

TABLE 12.6 Classification Report of ResNet-50

Classes	Precision	Recall	F1-Score	Support
0	0.83	0.85	0.84	34
1	1.00	1.00	1.00	1
2	0.60	0.67	0.63	18
3	0.88	0.88	0.88	32
4	0.90	0.72	0.80	25
5	0.61	0.65	0.63	17
6	0.92	0.96	0.94	25
Micro-average	0.81	0.81	0.81	152
Macro-average	0.82	0.82	0.82	152
Weight-average	0.82	0.81	0.81	152
Samples-average	0.81	0.81	0.81	152

12.5.4 DenseNet

In DenseNet, every layer receives extra input from previous layers and sends its feature mappings to the following levels. A CNN architecture called a DenseNet was created to improve feature reuse and address the vanishing gradient issue. DenseNets have dense connections between layers because every layer sends its feature maps to every layer that follows it and receives feature maps from every layer that comes before it. This structure of dense connectedness fosters feature reuse throughout the network, helps gradient propagation during training, and improves information flow. DenseNets are made up of dense blocks, each of which consists of several layers connected directly to one another. This allows for more effective parameter usage and better model performance. They are extensively employed in computer vision applications like segmentation, object identification, and image classification (Figures 12.9, 12.10, and Table 12.7).

12.5.5 Hybrid Model

For better depth learning, a hybrid model combining ResNet50's residual connections with VGG16's stacked convolutional layers would combine rich feature extraction. Attention mechanisms or feature map concatenation may be used in the fusion process. The model would be further refined and specialized for particular jobs, with a customized output layer. The hybrid model seeks to take advantage of the complementary capabilities of both architectures to potentially improve performance on a range of computer vision applications. Experimentation and fine-tuning are essential to maximize the performance of this combination (Figures 12.11, 12.12, and Table 12.8).

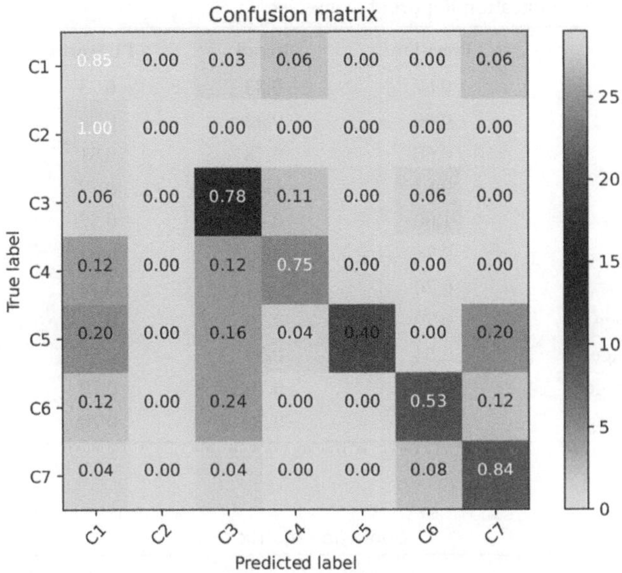

FIGURE 12.9 Confusion matrix of DenseNet.

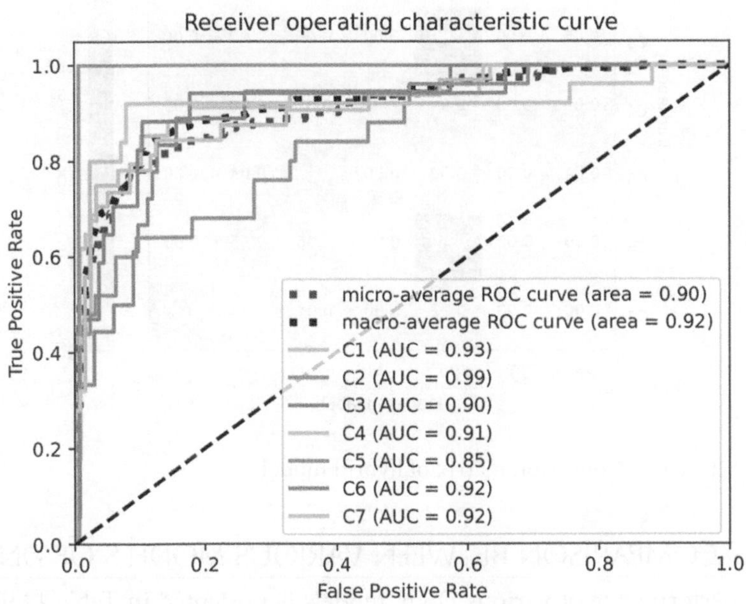

FIGURE 12.10 Receiver operating characteristic curve of DenseNet.

TABLE 12.7 Classification Report of DenseNet

Classes	Precision	Recall	F1-Score	Support
0	0.67	0.85	0.75	34
1	0.00	0.00	0.00	1
2	0.50	0.78	0.61	18
3	0.83	0.75	0.79	32
4	1.00	0.40	0.57	25
5	0.75	0.53	0.62	17
6	0.70	0.84	0.76	25
Micro-average	0.70	0.70	0.70	152
Macro-average	0.64	0.59	0.59	152
Weight-average	0.75	0.70	0.70	152
Samples-average	0.70	0.70	0.70	152

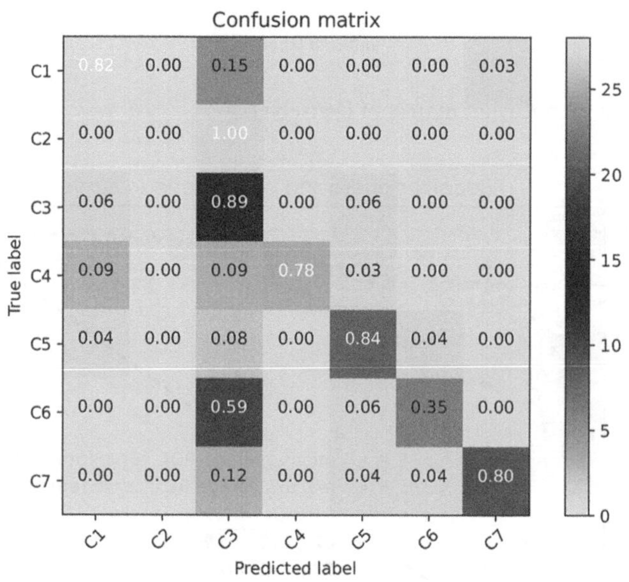

FIGURE 12.11 Confusion matrix of hybrid model.

12.6 COMPARISON BETWEEN VARIOUS MODELS OF DNN

The performance of various DNN models is presented in Table 12.9. The highest accuracy (95.21%) is achieved in the VGG-16 model. The second highest is achieved in the VGG-19. The hybrid model (Resnet-50 and VGG-16) performed better than Resnet-50 but not better than VGG-16.

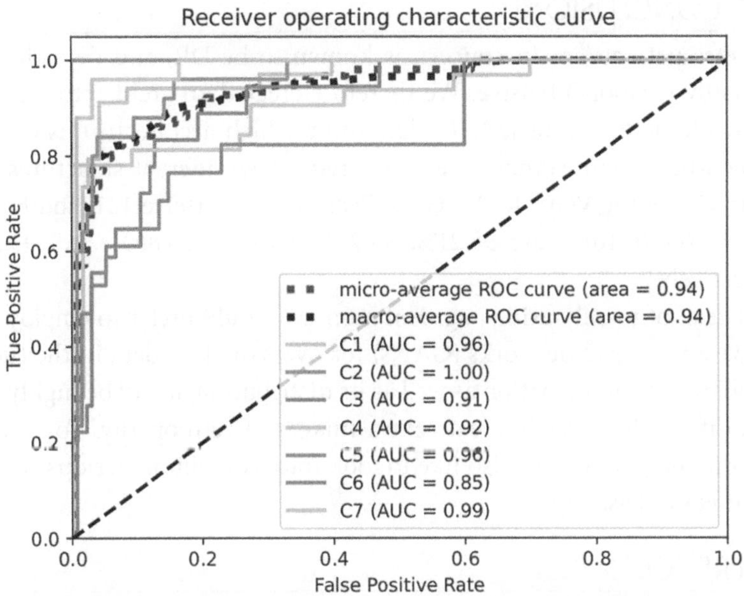

FIGURE 12.12 Receiver operating characteristic curve of hybrid model.

TABLE 12.8 Classification Report of Hybrid Model

Classes	Precision	Recall	F1-Score	Support
0	0.85	0.82	0.84	34
1	0.00	0.00	0.00	1
2	0.40	0.89	0.55	18
3	1.00	0.78	0.88	32
4	0.84	0.84	0.84	25
5	0.75	0.35	0.48	17
6	0.95	0.80	0.87	25
Micro-average	0.76	0.76	0.76	152
Macro-average	0.68	0.64	0.64	152
Weight-average	0.83	0.76	0.77	152
Samples-average	0.76	0.76	0.76	152

TABLE 12.9 Comparative Analysis of Different CNN Models

S. No	Model Name	No. of Epoch	Ratio Training and Testing Dataset	Accuracy (%)
1	VGG-16	10	80:20	95.21
2	VGG-19	10	80:20	94.21
3	Resnet-50	10	80:20	93.22
4	Densenet121	10	80:20	75.86
5	Hybrid model	10	80:20	93.55

12.7 CONCLUSION

Diabetes patients' main concern is known to be DR, and detecting DR manually is labor-intensive. We therefore created an architecture for the automatic detection of DR. To determine which architecture is optimal under what circumstances, we compared a few different structures. The accuracies of the VGG-16, VGG-19, ResNet-50, DenseNet121, and Hybrid model architectures are 95.21%, 94.21%, 93.22%, 75.86%, and 93.55%, respectively.

Instead of merely using augmentation, we would prefer to employ generative adversarial networks (GANs) for oversampling data in the future. To balance the information as well as its distribution, it will be highly beneficial if we deliberately create extra images of retinopathy. To enhance performance, we would also like to look into assembling various transfer learning models.

REFERENCES

1. Pirić, M., Šojat, D., Mišković, V., Šapina, M., Popović, Z., & Bačun, T. (2020). Glycaemia, renal function and body mass in men and women with type 2 diabetes. *Medicinski Glasnik (Zenica), 17*(2), 523–529.
2. Jenkins, A. J., Joglekar, M. V., Hardikar, A. A., Keech, A. C., O'Neal, D. N., & Januszewski, A. S. (2015). Biomarkers in diabetic retinopathy. *The Review of Diabetic Studies: RDS*, 12(1–2), 159.
3. Janghorbani, M., Jones, R. B., & Allison, S. P. (2000). Incidence of and risk factors for proliferative retinopathy and its association with blindness among diabetes clinic attenders. *Ophthalmic Epidemiology, 7*(4), 225–241.
4. Yau, J. W., Rogers, S. L., Kawasaki, R., Lamoureux, E. L., Kowalski, J. W., Bek, T., ... & Meta-Analysis for Eye Disease (META-EYE) Study Group. (2012). Global prevalence and major risk factors of diabetic retinopathy. Diabetes Care, *35*(3), 556–564.
5. Shaw, J. E., Sicree, R. A., & Zimmet, P. Z. (2010). Global estimates of the prevalence of diabetes for 2010 and 2030. *Diabetes Research and Clinical Practice, 87*(1), 4–14.
6. Yang, W., Lu, J., Weng, J., Jia, W., Ji, L., Xiao, J., ... & He, J. (2010). Prevalence of diabetes among men and women in China. *New England Journal of Medicine, 362*(12), 1090–1101.
7. Safi, H., Safi, S., Hafezi-Moghadam, A., & Ahmadieh, H. (2018). Early detection of diabetic retinopathy. *Survey of Ophthalmology, 63*(5), 601–608.
8. Network, S. I. G. (2010). *Management of Obesity: A National Clinical Guideline*. Scottish Intercollegiate Guidelines Network: Edinburgh, 20.
9. Bresnick, G. H., Mukamel, D. B., Dickinson, J. C., & Cole, D. R. (2000). A screening approach to the surveillance of patients with diabetes for the presence of vision-threatening retinopathy. *Ophthalmology, 107*(1), 19–24.

10. Hill, L., & Makaroff, L. E. (2016). Early detection and timely treatment can prevent or delay diabetic retinopathy. *Diabetes Research and Clinical Practice, 120,* 241–243.
11. Nguyen, T. T., Wang, J. J., Sharrett, A. R., Islam, F. A., Klein, R., Klein, B. E., … & Wong, T. Y. (2008). Relationship of retinal vascular caliber with diabetes and retinopathy: the Multi-Ethnic Study of Atherosclerosis (MESA). *Diabetes Care, 31*(3), 544–549.
12. Lechner, J., O'Leary, O. E., & Stitt, A. W. (2017). The pathology associated with diabetic retinopathy. *Vision Research, 139,* 7–14.
13. Kwan, C. C., & Fawzi, A. A. (2019). Imaging and biomarkers in diabetic macular edema and diabetic retinopathy. *Current Diabetes Reports, 19*(10), 95.
14. Raman, R., Ganesan, S., Pal, S. S., Kulothungan, V., & Sharma, T. (2014). Prevalence and risk factors for diabetic retinopathy in rural India. SankaraNethralaya Diabetic Retinopathy Epidemiology and Molecular Genetic Study III (SN-DREAMS III), report no 2. *BMJ Open Diabetes Research and Care, 2*(1), e000005.
15. Resnikoff, S., Felch, W., Gauthier, T. M., & Spivey, B. (2012). The number of ophthalmologists in practice and training worldwide: a growing gap despite more than 200 000 practitioners. *British Journal of Ophthalmology, 96*(6), 783–787.
16. Burgess, P. I., MacCormick, I. J. C., Harding, S. P., Bastawrous, A., Beare, N. A. V., & Garner, P. (2013). Epidemiology of diabetic retinopathy and maculopathy in Africa: a systematic review. *Diabetic Medicine, 30*(4), 399–412.
17. Sivaprasad, S., Gupta, B., Crosby-Nwaobi, R., & Evans, J. (2012). Prevalence of diabetic retinopathy in various ethnic groups: a worldwide perspective. *Survey of Ophthalmology, 57*(4), 347–370.
18. Grzybowski, A., Brona, P., Lim, G., Ruamviboonsuk, P., Tan, G. S., Abramoff, M., & Ting, D. S. (2020). Artificial intelligence for diabetic retinopathy screening: a review. *Eye, 34*(3), 451–460.
19. Badar, M., Haris, M., & Fatima, A. (2020). Application of deep learning for retinal image analysis: a review. *Computer Science Review, 35,* 100203.
20. Asiri, N., Hussain, M., Al Adel, F., &Alzaidi, N. (2019). Deep learning based computer-aided diagnosis systems for diabetic retinopathy: a survey. *Artificial Intelligence in Medicine, 99,* 101701.
21. Islam, M. M., Yang, H. C., Poly, T. N., Jian, W. S., & Li, Y. C. J. (2020). Deep learning algorithms for detection of diabetic retinopathy in retinal fundus photographs: a systematic review and meta-analysis. *Computer Methods and Programs in Biomedicine, 191,* 105320.
22. Nielsen, K. B., Lautrup, M. L., Andersen, J. K., Savarimuthu, T. R., & Grauslund, J. (2019). Deep learning–based algorithms in screening of diabetic retinopathy: a systematic review of diagnostic performance. *Ophthalmology Retina, 3*(4), 294–304.

23. Tsiknakis, N., Theodoropoulos, D., Manikis, G., Ktistakis, E., Boutsora, O., Berto, A., ... & Marias, K. (2021). Deep learning for diabetic retinopathy detection and classification based on fundus images: a review. *Computers in Biology and Medicine, 135*, 104599.
24. Stitt, A. W., O'Neill, C. L., O'Doherty, M. T., Archer, D. B., Gardiner, T. A., & Medina, R. J. (2011). Vascular stem cells and ischaemic retinopathies. *Progress in Retinal and Eye Research, 30*(3), 149–166.
25. Gulshan, V., Peng, L., Coram, M., Stumpe, M. C., Wu, D., Narayanaswamy, A., ... & Webster, D. R. (2016). Development and validation of a deep learning algorithm for detection of diabetic retinopathy in retinal fundus photographs. *JAMA, 316*(22), 2402–2410.
26. Xu, K., Feng, D., & Mi, H. (2017). Deep convolutional neural network-based early automated detection of diabetic retinopathy using fundus image. *Molecules, 22*(12), 2054.
27. Islam, S. M. S., Hasan, M. M., & Abdullah, S. (2018). Deep learning based early detection and grading of diabetic retinopathy using retinal fundus images. *arXiv preprint arXiv:1812.10595.*
28. Lam, C., Yi, D., Guo, M., & Lindsey, T. (2018). Automated detection of diabetic retinopathy using deep learning. *AMIA Summits on Translational Science Proceedings, 2018*, USA, 147.
29. Kassani, S. H., Kassani, P. H., Khazaeinezhad, R., Wesolowski, M. J., Schneider, K. A., & Deters, R. (2019, December). Diabetic retinopathy classification using a modified xception architecture. In *2019 IEEE International Symposium on Signal Processing and Information Technology (ISSPIT)*, United Arab Emirates (pp. 1–6). IEEE.
30. Jain, A., Jalui, A., Jasani, J., Lahoti, Y., & Karani, R. (2019, April). Deep learning for detection and severity classification of diabetic retinopathy. In *2019 1st International Conference on Innovations in Information and Communication Technology (ICIICT)*, Chennai, India(pp. 1–6). IEEE.
31. Tymchenko, B., Marchenko, P., & Spodarets, D. (2020). Deep learning approach to diabetic retinopathy detection. *arXiv preprint arXiv:2003.02261.*
32. Gangwar, A. K., & Ravi, V. (2021). Diabetic retinopathy detection using transfer learning and deep learning. In *Evolution in Computational Intelligence: Frontiers in Intelligent Computing: Theory and Applications (FICTA 2020)*, NIT Surathkal, India (Vol. 1, pp. 679–689). Springer Singapore.
33. Pratt, H., Coenen, F., Broadbent, D. M., Harding, S. P., & Zheng, Y. (2016). Convolutional neural networks for diabetic retinopathy. *Procedia Computer Science, 90*, 200–205.
34. Benítez, V. E. C., Matto, I. C., Román, J. C. M., Noguera, J. L. V., García-Torres, M., Ayala, J., ... & Grillo, S. A. (2021). Dataset from fundus images for the study of diabetic retinopathy. *Data in Brief, 36*, 107068.
35. Ain, Q. T., Ali, M., Riaz, A., Noureen, A., Kamran, M., Hayat, B., & Rehman, A. (2017). Sentiment analysis using deep learning techniques: a review. *International Journal of Advanced Computer Science and Applications, 8*(6), 41.

36. Mascarenhas, S., & Agarwal, M. (2021, November). A comparison between VGG16, VGG19 and ResNet50 architecture frameworks for Image Classification. In *2021 International Conference on Disruptive Technologies for Multi-disciplinary Research and Applications (CENTCON)*, Bengaluru, India (Vol. 1, pp. 96–99). IEEE.
37. Mandal, B., Okeukwu, A., & Theis, Y. (2021). Masked face recognition using resnet-50. *arXiv preprint arXiv:2104.08997.*

INDEX